"Sean Kelly has written a brilliant and singular book that shines a powerful light on our critical moment in history. To comprehend our time requires a larger frame of reference and deeper sources of insight than currently inform our cultural mindset. *Coming Home* brings both vision and scholarship, rigor and imagination, to this all-important task of historical self-reflection, allowing us to glimpse profound overarching patterns that make newly intelligible not only our history but the specific challenges of our Planetary Era.

"Kelly has achieved here what is perhaps the most creative multidisciplinary elaboration of Hegel's dialectic of history yet attempted, doing so by carefully integrating the critical contributions of Jung, of Teilhard de Chardin, of Morin, Grof, Campbell, and many others. With its publication, *Coming Home* immediately becomes necessary reading for our time—necessary, at least, for those in our precarious civilization who know it is their peculiar calling to try to grasp the big picture as accurately as possible."

RICHARD TARNAS, author of *The Passion of the Western Mind* and *Cosmos and Psyche*

"For those travelers seeking meaning in this crisis-riven time, here is both map and provisions for the journey. To give us our bearings, Sean Kelly sets the present perilous moment within a vast landscape of relevance, including the mythic and psycho-philosophic features of our shared history.... He provides the courage and beauty of his breathtaking scholarship, and bountiful meals with mystics, scientists, and other pilgrims. This is a work of rare distinction. Particularly arresting is the lens Kelly provides for viewing Time: its successive and overlapping spans, its accelerating cycles, and the Singularity of their convergence at this historical moment. I bow to Kelly's respect for the radical uncertainty facing us now. His work illumines not only the inescapable nature of this uncertainty, but also its capacity to awaken and ennoble us at this turning in our human journey."

JOANNA MACY, author of *World as Lover, World as Self*

"At the threshold of the Planetary Age that must, but may not, come, Kelly sketches an impossibly—but all the more possibly—Great Vision. It is not concocted of new-agey spiritual pontification. It unfolds in dazzling breadth and trusty erudition, laced by a wise and earthy uncertainty."

CATHERINE KELLER, Professor of Constructive Theology, Drew University; author of *Face of the Deep; On the Mystery*

"Sean Kelly takes as his call the entire sweep of Western thought. He brings order to its profusion and finds major meanings and directions overlooked in traditional histories of ideas, all the while highlighting fascinating and forgotten clues and revealing undertows whose significance is becoming increasingly relevant and urgent in our own time. There may be other ways of ordering this vast field, but the story that Kelly presents is elegant, economical, immensely insightful, and, above all, charged with hopeful possibilities for change."

FREYA MATHEWS, Associate Professor of Ecological Philosophy at La Trobe University, and coeditor of the journal *PAN* (Philosophy Activism Nature)

COMING HOME

THE BIRTH & TRANSFORMATION
of
THE PLANETARY ERA

SEAN M. KELLY

Lindisfarne Books • 2010

2010
Lindisfarne Books
610 Main St., Great Barrington, MA 01230
Lindisfarne Books is an imprint of Anthroposophic Press, Inc.
www.steinerbooks.org

Cover Image © by Dejan Novakov, Shutterstock Images
Cover and book design: William Jens Jensen

LIBRARY OF CONGRESS CATALOGING-IN-PUBLICATION DATA

Kelly, Sean M., 1957–
Coming home : the birth and transformation of the planetary era /
Sean M. Kelly.
    p. cm.
Includes bibliographical references.
ISBN 978-1-58420-072-7
1. Civilization, Western. 2. Civilization, Modern—21st century.
3. Social change. 4. Climatic changes—Social aspects.
5. Extinction (Biology)—Social aspects. 6. Geopolitics.
7. Sustainability—Social aspects. 8. Consciousness.
9. Counterculture. 10. World history—Philosophy. I. Title.
CB245.K43 2010
909'.09821—dc22

                                        2009048746

Printed in the United States

CONTENTS

Origin and goal are unknown to us, utterly unknown by any kind of knowledge. They can only be felt in the glimmer of ambiguous symbols. Our actual existence moves between these two poles; in philosophical reflection we may endeavour to draw closer to both origin and goal.

—KARL JASPERS

All descriptions of the past are in the present; therefore, history tells our descendants more about us than it does about the imaginary creatures we like to call our ancestors....

All of which is only another way of saying that the past and the future do not exist; nevertheless, we need these narrative fictions, for we gain knowledge by looking backward at patterns and forward in anticipation of the results of our actions.

—WILLIAM IRWIN THOMPSON

Wo gehn wir denn hin? "Immer nach Hause."
(Where then are we heading? "Ever homeward.")

—NOVALIS

INTRODUCTION

"For better and worse, we now occupy a human planet, one in which most evolutionary forces are guided or misguided by our hand.... Human agency will alter the fate of all living beings because no part of the planet is unaffected by our activity." —PAUL HAWKEN

FIVE centuries ago, following the great voyages of discovery, and of conquest, a new era began that involved an unparalleled increase and stabilization of communication and exchange between inhabitants of all of the world's continents. At the same time, thanks to Copernicus and his followers, European intellectuals started to accept the idea that the Earth, along with the other heavenly wanderers, is a planet (from the Greek, *planetes* = wanderer). Thus began the Planetary Era.[1] **The intercontinental exchange was material (gold, silver), biological (plants, animals, viruses), technological, and more broadly cultural.** Though obviously one-sided—an inevitable corollary of colonial domination—this communication and exchange has led to increasing economic and more generalized interdependence, to a growing sense of the complex human fabric that, however thin and prone to tearing, continues to weave itself around the planet. And though much thinner than the atmosphere or the rest of the biosphere, this new planetary layer is bringing about the end of another, geological, era—the Cenozoic—that has lasted for the last sixty-five million years.[*] The Mesozoic—along with its most famous inhabitants, the non-avian

---

[*] The three main ages of animal life on the planet are the Paleozoic: 543 million to 248 million years ago; the Mesozoic: 248 to 65 million years ago; and the Cenozoic: 65 million to the present.

dinosaurs—was brought to an end by the cosmic catastrophe of a massive meteorite smashing into the Earth and radically altering its climate. By contrast, the end of the Cenozoic—and with it the world's sixth mass extinction underway and accelerating in our time—is due entirely to human agency, especially since the beginning of the twentieth century with the deadly combination of the global population explosion and the unconstrained, empire and capital driven techno-industrial complex.[2]

Thomas Berry has suggested the term *Ecozoic* to describe the new age destined to follow the Cenozoic.[3] We shall have to wait to see how appropriate this proposed term turns out to be. The geological community, for its part, is debating whether or not to call this new age the *Anthropocene,* in recognition that humans are the principal drivers in bringing the Cenozoic to a close. There is no doubt, however, that the Earth community is in the grip of a series of overlapping and mutually reinforcing crises. There is, as a kind of persistent and deepening drone beneath the others, the ecological crisis represented by the destruction and possible collapse of the biosphere as it has existed for countless millennia. Most obvious at the time of writing are the global financial crisis and the chronic and regionally exacerbated sociopolitical crises. Though more entrenched in the world's hot spots or "fracture zones," like the Middle East, these crises are increasingly non-local, which is to say global, in their spheres of influence: witness the so-called war on terror and the global impact of the collapse of the American financial system.[4] Subtlest of all, and pervading the planetary anthroposphere like an invisible miasma, is a psycho-spiritual crisis of consciousness, a crisis of meaning and imagination. Our dominant habits of mind are not adequate to—and are arguably responsible for bringing about—the complex global polycrisis in which we find ourselves in this sixth century of the Planetary Era. We cannot—or do not wish to—see where we are heading. We have forgotten the path that has led us to where we now stand.

◎◎

I am aware of several potential challenges facing the sympathetic reception of the ideas presented in the following pages. To begin with, it requires a certain facility in multi-, inter-, and transdisciplinary modes of thinking.* In contrast to the dominant mode of over-specialization and disciplinary fragmentation, I draw freely from, and try to show the coherence of throughout, the full spectrum of the history of ideas: from philosophy, theology, and religious studies, from psychology (particularly depth and transpersonal), from the history and philosophy of science, and from political and cultural history. In this I have been guided by the example of such panoptic thinkers as C. G. Jung, Karl Löwith, Arthur Lovejoy, Jean Gebser, M. H. Abrams, William Irwin Thompson, Ken Wilber, Edgar Morin, and Richard Tarnas (key works of whom are listed in the References).

Related to this first, transdisciplinary challenge is the widespread suspicion in academic circles of grand (or "meta-") narratives, particularly when these assume or assert an intelligible pattern and intentional goal to the movement of history. The catastrophes of the twentieth century and the present planetary crisis certainly give the lie to any triumphalist or even unambiguously progressivist view of history or the evolution of consciousness. Though I assume no such view, I do discern a pattern and intuit a goal, the nature of which, however, I cannot articulate with the same degree of confidence or certainty as do Hegel, Aurobindo, and Teilhard, for instance, upon whose broad shoulders I am nevertheless forced to stand.

A third challenge concerns what can appear as a blatant occidentocentrism. If the main subject is the *Planetary Era*, how justify a focus

---

* The idea of *multi*-disciplinarity is self-evident. While *inter*-disciplinarity might be said to involve a focus on data, theories, or issues that overlap two or more disciplines, or as involving the application of models or methods from one discipline to another, *trans*-disciplinarity involves inquiry at the level of root or paradigmatic assumptions or principles that structure (usually unconsciously) any discipline.

on the West? The larger portion of humanity, after all, lives in the East and the South, and despite the dominance of the U.S., Europe, and the former Soviet Union in modern times, most analysts point to China and India as the global giants of the twenty-first century. Aside from the fact that, given my personal situatedness, a Western perspective is the only one directly available to me, it is the case that the imperialistic drive (discovery and conquest) that initiated the Planetary Era, as well as the scientific, industrial, and political revolutions that shaped the modern global landscape, all arose in the West. Hence the justification for an occidental focus in this treatment of the birth and transformation of the Planetary Era.

If I were to imagine a considerably expanded version of this book, I would include excursions on the vital contributions of the East—of China during the Enlightenment, for instance, of India and China during the Sixties Counterculture. And there would be much to say of the glories of Muslim culture during the Middle Ages and its critical role in mediating the development of the Western mind. My main purpose, however, has not been to provide an all-inclusive, planet-wide history of ideas, but instead to trace the evolution of the dominant Western worldview along with a particular stream of countercultural perspectives out of which it arose and with which, in fact, it has continued to be complexly related.

I propose that we can see the relation of the dominant to this countercultural stream of the Western worldview as an expression of a deep, dialectical/dialogical* and evolutionary pattern. This pattern can be described as a spiral embedded in an arc, or more precisely, a series of ever smaller arcs. The spiral reproduces the triphasic pulse of the arc (beginning, zenith, end), with each iteration happening in a shorter time span: a tightening spiral. The largest arc corresponds to the movement from the Alpha of human origins to an Omega that promises the possible stabilization of a truly planetary culture.

---

* The terms *dialectical* and *dialogical* are elaborated on in chapter 2.

The next shorter arc shares the same end but begins with the historical period. The third arc, always with the same end, begins with the Axial Period. First proposed by the philosopher Karl Jaspers, the Axial Period, in its principal phase, ranges from 800 to 200 B.C.E. In a manner resonant with the mysterious origin itself, this period saw the roughly simultaneous emergence of almost all of the world's great traditions of deep and abiding wisdom, including the first Greek philosophers (from Thales and Pythagoras to Plato and Aristotle), the Buddha, Mahavira, the writing of the Bhagavad Gita, Zoroaster, Confucius and Lao Tsu, and the great Jewish prophets (from Isaiah to Ezekiel).*

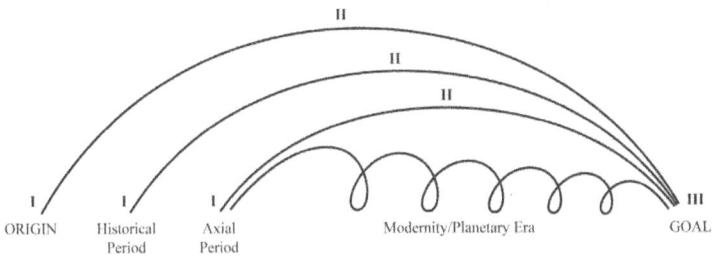

It is with a particular tracing of this third arc—one more or less coextensive with what Jung calls the *Christian Aion* (also the astrological age of Pisces)—that the bulk of the following pages is concerned. It begins during the last flowering of the Axial Period in the Middle East, which is to say between East and West, as also between North and South, with the origins of what will become the chrysalis (to use Toynbee's expression) of the dominant Western worldview until late modern times. Again, given the catalytic role of the West in the birth of the Planetary Era, I beg the reader's indulgence for perhaps seeming to privilege the largely Christian West over other cultural complexes. While I maintain the hermeneutical appropriateness of my focus (which only my extended argument can hope to bear

---

* For a more recent substantial treatment of the idea of Axial Age, see Karen Armstrong's *The Great Transformation: The Beginning of Our Religious Traditions*.

out), it should be obvious that I am in no way arguing for the cultural or spiritual superiority of the West.*

In fact, it is a special virtue of the dialectical quality of the pattern in question (the movement from a simple identity, through a differentiation, to a more complex identity) that it not only allows for, but actively encourages, an internal critique of the symbolic forms through which the pattern is enacted. These forms cut across the main lines of cultural expression, including religious doctrines, scientific theories, and political ideologies. My appeal later in these pages to Blake's notion of the Bible as the "Great Code" (the most relevant incarnation of the fundamental pattern for our purposes) therefore follows the visionary poet's example of underlining the critical distinction between the living core of the religious symbols and their dogmatically codified, and often co-opted, counterparts. The same holds for the analogous distinction between the nobly Promethean aspirations of modernity—notably, the ideal of freedom—and its shameful history of exploitation.

Both the diminishing span of the arcs and the tightening spiral point to the acceleration of the time sense that is so familiar and often disquieting to us as we age. This sense of acceleration, however, has become epidemic in Western, and increasingly global, culture at large. Does this mean that the culture is growing old and nearing its end? Many believe so, and even wish it to be so, given how unsustainable it has become. As we shall see in the later part of this book, there are compelling reasons to accept that we are indeed already upon the threshold that marks the end of the overlapping but increasingly shorter trajectories leading to our present moment: the historical period as a whole (5,000 years),† the period coinciding with the rise

---

* Writing from within what he calls the Christian saptiential tradition, Bruno Barnhart has worked out his own three-phase arch of the evolution of Western culture, which dovetails nicely with my proposal (see Barnhart, 2007, 102f.).

† By the end of the historical period, I am obviously not claiming that time is coming to an end, or that significant events will cease to transpire, or that histories will

of the West (1,000 years), modernity (500 years), the age of unchecked industrial growth (200 years).

Because they all involve a movement toward a zenith followed by a decline, these trajectories or arcs suggest the analogy of the apparent course of the Sun, which in many cultures has symbolized the mythic voyage of the hero. We shall consider this voyage in the first chapter. For the moment I would point out that the set of nested arcs (excluding the largest arc, from Origin to Goal) describes not the overall rise and fall of the solar principle as such, but more precisely this principle as it has manifested in association with the dominance of patriarchy throughout the historical period. During this period, the masculine principle has been traditionally associated with such qualities as competitive striving, independence or separativeness, dominance, and in later periods with a certain inflection of consciousness and rationality that stresses certainty, closure, and rigidly hierarchical thinking. If not necessarily expressive of masculinity as such, these qualities are widely recognized as typical of the masculine within patriarchal culture.

Can we then expect a rise or return of the "feminine"*—that is, a greater emphasis on such qualities as cooperation, relatedness, and

---

not longer be written. Neither am I necessarily invoking the Hegelian (or Kojèvian) notion of the end of history (see Kelly, 1993, especially pp. 170f.), and certainly not Fukuyama's neo-liberal flattening of this notion. Rather, I am simply pointing to the fact that what has constituted the main driving force and central theme of the historical process from its inception to the present—namely, the quest for autonomy and dominance on the part of individual nations—must now give way to a growing recognition of the fact of pervasive interdependence on a planetary scale (I consider this in more detail in chapter 12).

* I am aware of the complex, ever-shifting, and highly contested character of the debate surrounding questions of sex and gender. After several decades of increasingly well-received arguments supporting the idea of the social construction of what has been commonly understood by the terms *masculine* and *feminine,* recent studies of sex-linked differences in genetic profiles, brain anatomy, and neurochemistry have reinvigorated some of the more traditional naturalistic views. Evidence here of the proverbial pendulum swing should alert us to the danger of simplistic or reductionistic thinking on either side of the equation.

more embedded, holistic, fluid, and lateral thinking—on the other
side of the threshold? Given the extremity of our current situation,
we can only hope that this will be the case. We can at least say this
much about the path(s) covered thus far: with the initiation of each
new arc, and with each new turn of the spiral over the last two millennia, the countercultural impulse leading to the birth of a new phase of
consciousness and culture (the origins of Christianity during the late
flowering of the Axial Age, the birth of the modern, the Romantic–
Idealist movement, the Sixties Counterculture, to name the four most
well-known turns) drew from or emphasized such typically feminine
qualities as cosmic or organic embeddedness, universal sympathy, and
symbolic or analogical thinking. Though the dominant cultural context has been and remains solar–patriarchal, the periodic resurgence
of countercultural impulses can be seen to have aimed at an as yet
unconsummated *coniunctio* or marriage of the masculine and (both
repressed and oppressed) feminine principles. The evolution of the
Western mind, as Tarnas has noted,

> can be seen has having been marked at every stage by a complex
> interplay of the masculine and feminine, with significant partial
> reunions with the feminine having occurred in coincidence with
> the great watersheds of Western culture from the birth of Greek
> civilization onward. Each synthesis and birth has constituted a
> stage in the larger overarching dialectic between the masculine and
> feminine that I believe comprehends the history of the Western
> mind as a whole.[5]

Less sophisticated versions of this view of the masculine/femi-
nine dialectic are now quite widespread in spiritually inflected varieties of alternative culture. One of the major conduits of the view is
Jungian psychology, to core insights of which I shall have occasion to
appeal throughout these pages. This view has a much older pedigree,
however, stretching through the Romantics back to medieval mysticism (with Julian of Norwich, for example, who evoked God as both

Father and Mother) and the esoteric traditions (especially alchemy, with its vision of the androgyne as symbol of the philosopher's stone).*

If it is the case that, however unwittingly, the rise of the (solar-masculine) West initiated the Planetary Era, the survival, not only of a planetary humanity, but of the majority of the world's species now depends upon the full consummation of this (both longed-for and long-resisted) marriage.

The biblical mythos that has so deeply informed Western culture features not only the symbol of the sacred marriage (notably in both the Song of Songs and Revelation), but also that of the path or journey, which, after the Fall, is figured as a journey back home. This home is not the lost Garden, but the "New Heaven and New Earth" whose margins, as we have glimpsed them, have always receded as we move. *"Wo gehen wir denn hin?"* asks Novalis. *"Immer nach Hause."* ("Where then are we heading?" "Ever homeward.")[6†]

---

* In keeping with the alchemical vision of the *coniunctio*, Tarnas has more recently come to see this dialectic as intimately interwoven with that of another pair of archetypal principles—the solar and lunar—that, though overlapping in some respects with the masculine and feminine, remain nevertheless distinct. "I now consider the masculine-feminine dialectic," he writes, "as complexly intertwined with the solar–lunar dialectic, so that the solar masculine in patriarchy can be seen to have elevated itself at the expense of the lunar feminine (the caring mother and supportive wife), the solar feminine (the independent and assertive woman), and the lunar masculine (the relational, sensitive, sensuous, intuitive man)." Instead of the masculine and feminine, "the solar–lunar is the underlying dynamic at work in the hero's journey and the evolution of consciousness" (Richard Tarnas, personal communication. 12/2008).

† These lines are from the unfinished *Heinrich von Ofterdingen*. In his notes for the continuation, Novalis writes: "It is at the end the primal world, the golden age./ Men, beasts, plants, stones and stars, flames, tones, colors must at the end act and speak together as a single family or society, as a single race." (ibid., 252.) The receding margins from the preceding sentence is an echo of Tennyson's "Ulysses": "I am part of all that I have met; / Yet all experience is an arch wherethrough / Gleams that untravelled world, whose margin fades / For ever and for ever when I move."

PART ONE

FINDING OUR WAY

"Since the sixties and seventies, the future has become a dimension of our present. Like it or not, we must discern where history is moving. We have not yet developed sophisticated ways of doing this, but do it we must—willingly and humbly." —EWERT COUSINS

1

THE MONOMYTH

IN all cultures, questions of the origins, destiny, and essential identity of a people have always been answered in the form of a founding myth, which is to say a sacred narrative that gives symbolic expression to the fundamental nature of, and the relations between, the human, the cosmos, and the divine. Joseph Campbell proposed the term *monomyth** to describe the common pattern that he saw behind much of the world's mythology, a pattern he most often summed up with the notion of the hero's journey. Though he identified many possible elements of the hero's journey, he recognized three main steps or phases. Influenced by Van Gennep's theory of rites of passage (marking the transition through puberty, for instance), which all seem to involve (1) a separation from the day to day life of the given community, (2) a so-called *liminal* (from the Greek for "threshold") or transitional phase during which the initiation or transformation takes place, and (3) a reincorporation into the community, Campbell's three phases are: *separation, initiation,* and *return.* Here is Campbell's original formulation of the hero's journey: "*A hero ventures forth from the world of common day into a region of supernatural wonder: fabulous forces are there encountered and a*

---

* A term that he borrowed from a passage in James Joyce's *Finnegans Wake:* "When they were all there now, matinmarked for lookin on. At the carryfour with awlus plawshus, their happyass cloudious! And then and too the trivials! And their bivouac! And his monomyth! Ah ho! Say no more about it! I'm sorry! I saw. I'm sorry! I'm sorry to say I saw!" (Joyce, 1999, 581).

decisive victory is won: the hero comes back from his mysterious adven-
ture with the power to bestow boons on his fellow man." [7]

Although this formulation certainly fits a large sampling of the
hero's journey, I would question the necessity of some kind of "decisive victory" and the "power to bestow boons." As we shall consider
in a moment, the initiatory phase often highlights a defeat (death,
suffering, trials), the passage through which may or may not be con-
strued as a victory. Sometimes it is a question of mere survival that
comes at the cost of distinct failure. Similarly, though there is generally some kind of gain upon the return, if only in form of the wisdom
of experience, the boons can be ambiguous and are often offset by
what amounts to tragic losses. There is generally still, however, the
sense that the journey was somehow necessary or in some sense profoundly valuable, that we who hear and retell the stories have something to learn, and that we forget them at our peril.

Among the hundreds of possible illustrations one could turn to,
I would comment briefly on the following myths, all of which have
served as guiding narratives for whole societies or widely diffused
cultural groups: the story of Gilgamesh, from the oldest written
epic; the Homeric wanderings of Odysseus; the Woman Who Fell
from the Sky, or the so-called Iroquois cosmology; the Valentinian
Gnostic myth of Sophia; and for a contemporary and popular literary example, the quest of Frodo Baggins from Tolkien's *The Lord of
the Rings*. These illustrations will not only help us get a more concrete sense for the structure of the monomyth, but will also serve to
amplify the more familiar biblical version that will inform much of
the subsequent discussion.

The epic of Gilgamesh precedes the Axial Period by around
two millennia. It is the earliest known text accompanied by an
author's name, Shin-eqi-unninni. It also portrays one of the first
great if tragic personalities in the early age of Empire, and perhaps
has something to say about Empire's deeper origins and eventual

fate. Gilgamesh is two-thirds divine and one-third human, one of the first superheroes whose exploits are said to have been carved in lapis lazuli by the gates of the ancient city of Uruk. At the beginning of the account, we read that:

> He saw the great Mystery, he knew the Hidden:
> He recovered the knowledge of all the times before the Flood.
> He journeyed beyond the distant, he journeyed beyond
> exhaustion,
> And then carved his story on stone.

A major focus of this epic is the relationship between Gilgamesh and his initially wild counterpart or "brother," Enkidu, who was created by the gods to distract Gilgamesh from his puerile and oppressive behavior toward his subjects. There are two main journeys or quests in the epic. The first involves both Gilgamesh and Enkidu, who decide out of sheer wantonness to cut down the beautiful cedar forest in southern Iran, including the single giant tree—the World Tree, or *axis mundi*, in this case a symbol of archaic or original harmony with the ways of the cosmos—and also to subdue the spirit guardian of the forest Humbaba. In revenge for their destructive acts-, including the slaying of the Bull of Heaven and the insulting of the Great Goddess Ishtar, Enkidu is made sick and, after suffering terribly for twelve days, finally dies. This throws Gilgamesh into an agitated depression, not only out of grief for his lost friend but also, and especially, because he now realizes that he too is mortal. He therefore decides to set out on the second, solo, quest, to find Utnapishtim, the Babylonian Noah, who survived the Flood and holds the secret of immortality.

Gilgamesh voyages to the ends of the Earth and finally meets Utnapishtim, who gives him the task of staying awake for six days and seven nights, but Gilgamesh no sooner sits down than he falls fast asleep, losing his chance at immortality. "O woe!," he cries, "What do I do now, where do I go now?"

Death has devoured my body,
Death dwells in my body,
Wherever I go, wherever I look, there stands Death!

As a consolation, Utnapishtim tells him of the plant at the bottom of the sea that will make him young again. Gilgamesh successfully retrieves the plant, but on his way home as he is sleeping, a snake eats it (which is why snakes now shed their skins) and he is left empty-handed. Once again despondent, he laments: "For whom have I labored? For whom have I journeyed? For whom have I suffered?"

The tale ends with Gilgamesh back in front of the gates of Uruk, rather hollowly boasting of the city's greatness and pointing to the stone of lapis lazuli that tells of his wondrous exploits.

The first quest in this myth is emblematic of the patriarchal and inflatedly anthropocentric character of historical, urban-agrarian (and obviously also later industrial) civilization. The felling of the cedar forest and its majestic World Tree is an unambiguous expression of the antagonistic relation of historical civilization to the natural world and the ways of the cosmos, an antagonism echoed in the denigration of Ishtar (the mythic prototype here is the slaying of Tiamat by Marduk as recounted in the Babylonian epic of creation).[8] The second quest reveals the deeper motivation of the first—namely, a flight from finitude, from embodiedness and its inevitable confrontation with old age and death. As we have seen, the structure of the monomyth is patterned off of rites of initiation, and in this case especially the passage of young males into adulthood. This passage typically involves a kind of near-death experience, during which the individual not only "puts aside childish things" but also is initiated into the culture's meaning-bestowing central symbols, which in traditional (pre-urban) societies include a sense of alignment or harmony with the ways of the cosmos. Though clearly a compelling expression of the emergence of the solar-masculine hero, from our perspective the story of Gilgamesh, with its alienation of consciousness from its life-sustaining matrix, must be seen as a first or partial initiation only.

Though both are attributed to Homer, the *Iliad* and the *Odyssey* recount the stories of two different kinds of heroes. Achilles, the hero of the *Iliad*, embodies the traditional warrior virtues (*arete*) of honor and courage in battle—though he is also prone to excessive and destructive rage. As a result, Achilles never makes it back home. Odysseus does make it home to Ithaca to rejoin his faithful wife Penelope who, through a cunning to match that of her absent husband, has managed to keep her oppressive suitors at bay for years. According to Julian Jaynes,[9] the cunning of Odysseus is an expression of a more general mutation of consciousness in the direction of greater interiority and self-reflexivity, evidenced in the shift (from the *Iliad* to the *Odyssey*) in the meaning of such words as *nous* and *psyche*, which originally meant "gaze" and "breath," respectively, toward what we understand by the notions of mind and soul. Campbell, for his part, sees in the *Odyssey* an embodiment not only of the hero's journey but also of the *hieros gamos*, the archetype of the sacred marriage. Odysseus arrives home on the last day of the nineteenth year away, when the new moon coincided with the new sun of Winter Solstice, a day that was called the "Meeting of the Sun and Moon."[10] His initiation during the extended liminal phase of his wanderings—the descent of the solar hero—takes place through his encounters with Circe, Calypso, and finally Nausicaa, at whose father's court Odysseus recollects his journey up to his present circumstances, this final threshold of his homecoming. As Campbell notes, Odysseus "goes under the world in the West, visits the realm of the dead, and comes up in the extreme East, 'where the Daughter of Dawn has her dwellings and her dancing floors and the sun is uprising'—whereas Penelope, as we all know, was sitting at home, weaving a web and unweaving, like the moon."[11]

Despite the archetypal richness of journey, the tale ends in a general massacre, not only of the suitors and most of their kinsmen, but of the "faithless" maids as well, who, because they had slept with the suitors, are forced to carry out the dead bodies, and clean up after the carnage. In the end, "their heads were all in a line, and each had her neck caught fast in a noose, so that their death would be most pitiful."[12] Though he has perhaps come home a wiser man, he is still preoccupied with restoring his honor and reestablishing his place as lord and patriarch.

Turning now from pre-Axial Mesopotamia and Greece to indigenous North America, we find the striking tale of an Iroquois cosmological Eve—the Woman who Fell from the Sky. Before her "fall," Sky Woman is the heroine of her heavenly lodge (associated with the Moon). Prompted by the spirit of her deceased father, she embarks on a quest to the lodge of the sun chief (separation). During her initial visit, she passes several painful trials put to her by the chief (initiation)—including being harassed by dogs and being scalded by boiling water while preparing corn mush—and as a result is able to return with the boon of maize and venison. Following this successful quest, she returns to marry the sun chief, who soon becomes mysteriously ill and suspicious of his pregnant bride. In the longest version of the myth, he has the great sun-blossom bearing Tree near the lodge uprooted, and after luring her to the edge of the gaping hole in the sky, pushes her over and into her fall, thereby setting in motion a second and more momentous separation—of Earth and Sky.

The Woman who Fell from the Sky is assisted in her descent by circling ducks and the sacrificial acts of Beaver, Otter, and Muskrat, who dive to the bottom of the ocean to retrieve enough dirt to grow the Earth on the welcoming back of the Great Turtle. After her safe landing, and once the Earth has grown into a kind of Eden, she gives birth to her daughter, who quickly grows to marrying age. The daughter is wooed by a stranger and becomes pregnant with twins (as with her mother, the impregnation is "virginal," in this case occasioned by the stranger simply leaving one of his arrows by her as she sleeps). Less fortunate than her mother, the daughter dies in childbirth as a result of the evil brother, "Flint," who decides to exit through her armpit instead of through the vagina , the path taken by the good brother, "Sapling."

Flint lies to his grandmother, saying it was Sapling who killed her. Sapling is therefore cast away (separation), and after a period of solitude is initiated by his Spirit–Father (none other than the Great Turtle) at the bottom of a lake. As a result, he is now possessed of the drive and power to increase the Earth, to create populations of animals and the first human beings. The rest of the myth centers on the conflictual but nevertheless generative relation between the warring brothers, whose actions give final shape to the world. After a period of peaceful coexistence, Flint is killed in a final struggle. Sapling continues to wander the creation and encounters two males who, through their enlightenment at the hands of Sapling, become the first ancestors of the male initiation rites.

As in the case of Gilgamesh and Odysseus, I have only related the bare outline of this exceptionally rich mythic cycle, and I would refer the interested reader to the extended version as well as to Campbell's brilliant and extensive commentary in his *Historical Atlas of World Mythology* (II, 2). Here I would point out the following: in contrast with the more patriarchal myths of Gilgamesh and Odysseus, it is the questing woman who enacts the ideal (heavenly) prototype of the heroic quest. The theme of the sacred marriage of Sun and Moon is more explicit (though less symbolically elaborated) than in the *Odyssey*, and the Sky Woman's initiation and descent is the action that leads to the eventual creation of the world as we know it. The World Tree is uprooted, though not felled as in the myth of Gilgamesh, and only temporarily to open the way for the downward, incarnational movement of creation. The Tree is set right again once the heroine, pregnant with the future mother of the primal Twins, has passed through the heavenly portal. It is true that the role of creator is taken over by her son, Sapling.

Despite his special relation to his Spirit-Father, however—or perhaps
because of his successful initiation by the Spirit-Father—Sapling maintains a respectful and affirmative relationship with the Grandmother.
Similarly, rather than glorying in cutting down sacred forests and
slaughtering animals, he revels in bringing forth life. His actions
embody a masculine path in harmony with the ways of the cosmos.

Back now to the Old World, during the late flowering of the
Axial Period around the time of the origins of Christianity, we find
another compelling myth of a cosmogonic feminine figure in the
Gnostic Valentinian account of the trials of Sophia. From one of the
more complete and coherent accounts, we are told that, in the beginning, alone among the community of Aeons or divine emanations surrounding the mysterious Gnostic Godhead, Sophia persisted in her
longing to know the "ineffable Greatness" of the Father. She is pre-
vented from doing so through the actions of a certain "Limit," who
not only restores her to her twin brother emanation, "Willed," but also
separates her from her tormenting passions and casts them outside the
Pleroma or heavenly economy. These passions, which are described as a
kind of abortion, become the "lower" Sophia who, because of her exile,
undergoes her own sufferings. Already we can note a parallel with the
Iroquois cosmology. In both cases we have a divine feminine figure
who suffers and is cast out of Heaven (as both Eve and Adam, in the
more familiar biblical narrative, are cast out of the garden). In this case,
however, Sophia has not undergone a prior initiation.

In response to the sufferings of the lower Sophia, all of the Aeons
together produce a single Aeon, Jesus—"the perfect fruit of the
Pleroma"—who separates her from her passions and imparts to her
a preliminary gnosis or saving insight. While this does not yet bring
her back into the Pleroma, it nevertheless functions as a kind of initial return. The passions now become the prime matter of the cosmos.
Out of the "psychic" element she creates the "Demiurge" or inferior
creator god who, in complete ignorance of his origins, fashions the

cosmos and the first humans. The light the lower Sophia absorbed from the angels surrounding Jesus becomes the "pneumatic" or spiritual element, which she will eventually secretly introduce into the Demiurge's creation.

In contrast to the world of the Iroquois cosmology, whether before or after the Sky Woman's fall, the Gnostic cosmos is generally portrayed as inherently evil and deficient, and in this case the result of ignorance and boundless longing. Whereas in the Iroquois cosmology, the Fall into creation follows the sacred marriage of the lunar Sky Woman and the solar Chief, in the Valentinian system it is only at the end of time that the lower Sophia is united with Jesus (as the souls of the elect are united with their angelic counterparts surrounding Jesus) in the "bridal chamber" of the Pleroma. At this point, the final gnosis is realized, and with it, the whole cosmos evaporates into its original nothingness (ignorance).

Again, I have only touched on some of the main elements of this complex myth. While I agree with those who, like Elaine Pagels and many Jungians, appreciate the prominence accorded the feminine divine and the imaginal richness of the Gnostic myth of Sophia, this appreciation must be tempered with a recognition of what Hans Jonas terms Gnostic *acosmism*—that is, an uncompromising flight from the cosmos and an associated denigration of the body and materiality in general. Though there were other reasons as well, it was largely in reaction to this acosmism that the early Church Fathers, despite the patriarchal values they otherwise espoused, condemned the Gnostic myths. In my estimation, this is also why Jung, after his early enthusiasm for the Gnostics, eventually abandoned them in favor of the alchemical tradition, which tends to see the cosmos as a temple instead of a tomb or prison, and matter as sacred and involved in a process of ongoing transformation, rather than as demonic and destined to reveal itself as the mere privation of knowledge and being that it truly is.

We now jump ahead two millennia to our own time to consider Tolkien's *The Lord of the Rings*, a grand mythic epic that, especially after the major movie productions, has caught the imagination of hundreds of millions around the world. (The book was repeatedly voted the most popular of the twentieth century.) Tolkien's world is more vast than that of Homer or even the Valentinians and includes accounts of the Creator God (Ilúvatar), angelic intermediaries (the Valar and Maiar), and the actions of the Elves and Men (that is, humans) in the primordial First and Second Ages. The main action of *The Lord of the Rings* takes place at the end of the Third Age, which has seen the gradual resurgence of the evil power of Sauron (an original Maia who, like the biblical Lucifer, rebelled against Ilúvatar and the Valar, and was defeated but not destroyed at the end of the Second Age). This Third Age has also seen the rise and fall of the great Númenorean kings— the analogues of our mythic human (if sometimes, as with Gilgamesh or Achilles, also partly divine) heroes—and the gradual fading of the Elves and, more generally, of the enchanted phase of Middle-earth. Though imagined as occurring in our own distant past, this period, with its global struggle between the forces of death and domination, on the one hand, and those of life and freedom, on the other, resonate profoundly with our own time of planetary crisis.

Just as the adventures of the Iroquois hero Sapling are prefigured by the quest of Sky Woman before her "fall," so Frodo Baggins's quest in the *The Lord of the Rings* is preceded by his uncle Bilbo's "there and back again" journey, chronicled in *The Hobbit*, in which Bilbo sets out to retrieve the treasure-horde guarded by the dragon Smaug. It is upon his return to the idyllic Shire that Bilbo encounters the pitiful creature Gollum and finds the long lost Ring of power, the "one Ring to rule them all...and in the darkness bind them" that Sauron had created in his bid for total domination of Middle-earth. Under the guidance of the wizard Gandalf, Frodo is chosen as the new ring bearer, whose task it is, with the help of the fellowship of the

Ring, and especially his faithful companion, Sam, to make his way to Sauron's wasteland kingdom, Mordor, and cast the Ring into the fiery pits of Mount Doom in which it was originally forged.

Following his separation from his garden-like home in the Shire, Frodo undergoes an extended initiation, coming close to death a number of times, first at the hands of demonic Ringwraiths, then during his encounter with the giant spider, Shelob, and finally as he approaches the end of his quest, exhausted and inwardly poisoned by the burden of the Ring. From one perspective, Frodo fails in his quest as, poised above the Crack of Doom, he is overcome in his exhaustion by the power of the Ring and claims it for himself, only to have it bitten from his hand in his final struggle with Gollum, who tumbles to his death clutching his "precious" and so inadvertently brings the quest to a successful conclusion. After Frodo and Sam are rescued from Mount Doom by the Great Eagle, Landroval, and have recovered from their trials, there is the wedding of Aragorn—last in the line descended from the Númenorean kings—and the half-elf, Arwen Evenstar who, in choosing to remain in Middle-earth with her beloved, gives up her immortality.

Along with the basic structure of separation, initiation, and return, *The Lord of the Rings* shares other features with the previous myths, though again the differences are instructive. If Aragorn, the warrior King returned and wed to Arwen, resonates with Odysseus reunited with Penelope, Aragorn embodies superior virtues, including not only courage, cunning, and perseverance, but also a truly kingly devotion to the flourishing of the entire community of Middle-earth. His marriage to Arwen is paralleled by an inner marriage of the masculine and feminine (and solar and lunar principles), which we see in his wisdom and compassion and his ability to heal the sick. Frodo, for his part, though no less heroic than Aragorn, is in many ways the antithesis to Gilgamesh. In contrast to the inflated, narcissistic, and reckless Babylonian, the "Halfling" Frodo is humble and reserved. He is motivated in his quest not by longing (for immortality) or fear (of death), but by love and his devotion, as with Aragorn, to the greater good of the Middle-earth community. Sam, likewise, is not wild or only superficially tamed, like Enkidu, though he lives in harmony with the spirits of the Earth. Along with his green thumb and culinary gifts, we are told that, following his return to Hobbiton and his marriage to Rose, he re-greened the Shire after its devastation at the hands of the fallen wizard, Saruman. This he accomplished with the help of magical Earth—a gift from the Elvin Queen, Galadriel—which caused the saplings he planted to grow at an accelerated rate. One is reminded of the power of earth-increase of the Iroquois hero, Sapling.

The last lines of *The Lord of the Rings* are uttered by Sam. He is returning to the Shire from the Grey Havens, from which Frodo, Bilbo, and Gandalf have sailed to the undying lands in the far West.

> Sam turned to Bywater, and so came back up the Hill, as day was ending once more. And he went on, and there was yellow light, and fire within; and the evening meal was ready, and he was expected. And Rose drew him in, and set him in his chair, and put little Elanor upon his lap.
> He drew a deep breath. "Well, I'm back," he said.[13]

Again, if Rose is Sam's Penelope, and despite the stereotypical family scene with which *The Lord of the Ring* closes, the ideal of the hero that Sam embodies is as far as possible (without becoming that of the antihero) from the earliest Western prototypes of Gilgamesh, Achilles, and Odysseus. Though set in an imagined pre-history, Sam's homecoming can be seen as symbolic of our own, which, however, despite whatever glimpses of a yellow light and a fire within, is by no means assured.

2

A MORE FUNDAMENTAL PATTERN

CAMPBELL'S formulation of the monomyth conceived as the three phases of the hero's journey is fruitful and illuminating. There is a more general and in some ways more fundamental pattern, however, of which the hero's journey is certainly one of the most compelling manifestations. The most fully developed articulation of this pattern was given by G. W. F. Hegel (1770–1831), who saw himself as having translated the meaning of the great myths and reflections on the nature of God into a comprehensive system organized around the intellectual intuition of the Absolute or the Whole. This system was accompanied by a truly transdisciplinary logic or method that could serve not only to organize all existing realms of knowledge, but also to lift the veil from the Wisdom that philosophers had always sought.[14]

Hegel's logic of the Absolute is perhaps best known through the idea of the *dialectic* as the movement which leads from an initial position through opposition or negation to a new position that includes (or, to be more precise, simultaneously transcends, includes, and negates) the initial position along with its apparent opposite or "antithesis."* Hegel associates several triads of terms with the logic of the Absolute—most notably: the Idea (or Logic), Nature, and Spirit

---

* The word Hegel uses for "transcend and include" is *aufheben*, usually translated as "sublation." I put "antithesis" in quotation marks because Hegel himself actually never uses the formula (now so firmly associated with his name) of thesis, antithesis, synthesis. The formula is, however, used by his older contemporary, Fichte.

(a more traditional analogue here would be God, Cosmos, and the Human); universal, particular, and individual; essence, appearance, and actuality. For our purposes, the most helpful expression of the logic of the Absolute is the triad: *identity, difference,* and *new identity.*[*]

That this triad represents a more fundamental pattern is evident, to begin with, in the fact that Campbell's sequence clearly implies a prior identity as the base from which to separate. Campbell's first phase—separation—can therefore be considered an expression of the moment of difference in the more fundamental pattern. Initiation corresponds to the process leading from mere difference to a new identity (which is therefore not a "return" in the literal sense).

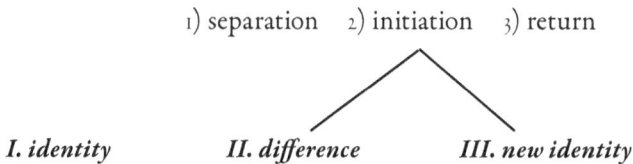

1) separation   2) initiation   3) return

I. identity   II. difference   III. new identity

As we shall see in a variety of contexts, the manner in which the third moment of new identity involves some kind of return is complex. Though the protagonist of the monomyth usually retains the same name, for instance, the second moment typically involves some kind of initiation, or at least a profound enough transformation to allow for the emergence of the new identity. And not only is the protagonist transformed, but also generally the place from which he or she

---

* The actual triad proposed by Hegel in the *Logic* is identity, difference, and ground. "Ground" (*Grund*) here could also be translated as "reason," particularly as in the philosophical notion of "sufficient reason" in the sense of the expression "raison d'être." "The *Ground*," writes Hegel, "is the unity of identity and difference, the truth of what difference and identity have turned out to be—the reflection-into-self [that is, the self-reflexivity of identity], which is equally a reflection-into-other [that is, the relationality of difference], and vice-versa. It is essence [or the idea of something's true or essential nature] put explicitly as a totality" (Hegel, 1975, section 121). Hegel's early formulation of the nature of the Absolute or the Whole was as "the identity of identity and difference (or non-identity)" (see Hegel, 1978).

set forth: Sky Woman's village now has maize; Middle Earth sees the departure of the ring bearers and their enchantments.

The Hegelian triad can also be seen as the deep structure of other—and arguably all—developmental processes. In my previous work,[15] I have shown how this is the case for Jung's understanding of the process of individuation or the actualization of the Self, where the three moments are: 1) the original "participation mystique" or identification with the archetypes of the collective *unconscious*.* Individually, this manifests as identification with the mother and the wider surround and is associated with archaic, magical, and mythic cognition as well as an original proximity to the numinous realms of spirit; 2) the differentiation of the individual *ego* (or separate self sense), especially the mental ego and its conceptual, or formal-operational, cognition, and its relation to the inner "other" (the shadow and the anima/animus), which tends to be projected and so colors external relations; and 3) the conscious emergence of the (true) *Self* as the (deeper) center and circumference of the psyche in its totality. This Self, though present from the beginning as archetypal possibility, is actualized through a recuperation of pre- or non-egoic potentials and an integration of the inner other(s). The individuation or actualization of the Self is associated with deepening into what Jung calls the symbolic life and manifests as a progressive embodiment of wholeness, a healing of developmental lesions, and a more conscious and wholesome relationship to the holy or sacred.

A finely nuanced version of this same triphasic pattern can be found in the works of Owen Barfield, one of the lesser-known "Inklings" that included Tolkien and C. S. Lewis, upon both of whom his thinking had an appreciable influence. Barfield drew from the

---

* I should stress here that this moment is "unconscious" only from the point of view of the next, or second, moment. Jung himself came to characterize the collective unconscious as a "multiple consciousness" (Jung, C.W. 8: 388ff.).

same source traditions as Jung (especially Goethe and the nineteenth
century Romantic tradition), and though friendly toward Jung's
thinking, considered himself more a close student of Rudolf Steiner.
For Barfield, the evolution consciousness proceeds through the initial
phase of "original participation," through the phase of "idolatry"—by
which he means the collapsed imagination of materialistic positivism—to the third phase of "final participation." "The elimination of
original participation," he writes,

> involves a contraction of human consciousness from periphery to
> centre...a contraction from the cosmos of wisdom to something
> like a purely brain activity—but by the same token it involves an
> *awakening.* For we awake, out of universal—into self—conscious-
> ness. Now a process of awakening can be retrospectively surveyed
> by the sleeper only after his awakening is complete; for only then
> is he free enough of his dreams to look back on and interpret them.
> Thus, the possibility to look over the history of the world and
> achieve a full, waking picture of his own gradual emergence from
> original participation, really only arose for man with the culmina-
> tion of idolatry in the nineteenth century. He has not yet learned
> to make use of it.[16]

We shall consider in greater detail the nature of this awakening and its
relation to the movement of history in the next and following chapters.

Variations of the same threefold pattern are also common in
more recent transpersonal or integral developmental models (of
which I shall have more to say in chapter 11): Ken Wilber's[17] understanding of the movement from the prepersonal, through the personal, to the transpersonal; Stanislav Grof's[18] theory of the perinatal
process (from the initial, "oceanic" identity of the fetus, through the
birth struggle, to the identity of the neonate)*; Michael Washburn's
pre-egoic, egoic, and trans-egoic;[19] and Jean Gebser's preperspectival

---

* Tarnas (1991, 430) was the first to make the connection between Grofian perinatal dynamics and the Hegelian dialectic. I have elaborated on this connection in
"James, Grof, and the Varieties of Perinatal Experience" (Kelly, forthcoming).

(archaic, magical, and mythic), perspectival (mental), and aperspectival (integral).[20]*

In terms of three of the myths considered in the previous chapter, we see the first moment symbolized in the Gnostic Pleroma prior to Sophia's sufferings and in the Iroquois Heaven before the Sky Woman's fall. Though echoed in the Eden-like portrayal of the Shire before the identity of the Ring of power is revealed, the deeper first moment in Tolkien's world corresponds to the First Age, before Sauron's rebellion and the fall of Númenor. In all three cases, the moment of difference involves some kind of fall and an active turning away from Source and a consequent loss of the state of original harmony and perfection. This moment is also associated with an intensification of self-reflective awareness and the clear differentiation of the moral opposites of good and evil: Sapling and Flint; Sophia and the Demiurge; Ilúvatar and Sauron.

The third moment, as we have seen, is often associated with the symbols of marriage and homecoming. Both depend upon the successful completion of the quest. Though Gilgamesh fails in his quest

---

* Examples could be multiplied. Two more from the field of speculative, and what could be called "integral" philosophy: 1) Sri Aurobindo's grand ontological trinity of the Ignorance or Inconscient, the lower hemisphere of Life and Mind, and the upper hemisphere of Supermind; his triphasic "logic of the Infinite" (Aurobindo, 1951, 295ff.), which involves the dialectical relation between the cosmic, the individual, and the Absolute; and the related "Supramental time consciousness," which involves the relation between the "timeless eternity," the "time movement," and the "simultaneous integrality of time" (ibid., 328). Related to Aurobindo's Supramental time consciousness is 2) Whitehead's cosmological "threefold creative act" of the universe (Whitehead, 1978, 525/346): the first moment of this act is "the one infinite conceptual realization," by which I understand the infinite potential of the "primordial nature" of God. This corresponds nicely to Hegel's Idea and to Aurobindo's first moment of "timeless eternity." Whitehead's second moment is "the multiple solidarity of free physical realizations in the temporal world"—a clear match with Aurobindo's "Time movement." The third is "the ultimate unity of the multiplicity of actual fact with the primordial conceptual fact," which again is a very good match with my suggestion for Aurobindo's third moment, the "simultaneous integrality of Time." For a discussion of these concepts relative to the question of the soul and the notion of what I call "integral time," see Kelly 2008b.

for immortality, judging from the opening lines of the epic, a certain
gnosis or wisdom is glimpsed ("He saw the great Mystery, he knew
the Hidden") through contact with the potentials corresponding
to the first moment. ("He recovered the knowledge of all the times
before the Flood.") Following the sack of Troy, Odysseus's quest
is, on the surface, the simple desire for the homecoming as marital
reunion, though the wisdom of experience, and the experience of
wisdom (Campbell's reading of a subtler motive of initiation through
the feminine), are also suggested. Sophia's quest in the Valentinian
Gnostic myth is to reclaim the light or divine sparks of Spirit that
she secreted in the cosmos (it is the gaining of gnosis, or the specially
revealed insight, which effects this reclaiming). There is healing in
the end with her marriage to Christos, but it comes at the cost of the
cosmos. In contrast with Sophia, the Iroquois Sky Woman commu-
nicates directly with the spirit of her father and undergoes a successful initiation and marriage in Heaven (= the *Pleroma*) prior to her
descent, which sets in motion a positively valued creation. Her grand-
son Sapling's quest is initiation by the father. Afterward, he manifests
the cosmos-positive power of earth-increase.

This power, as we have seen, is shared by Sam in *The Lord of the
Rings*. Sam's quest begins as an expression of devotion to Frodo, but
it matures (without diminishment of his devotion) into full participation with Frodo's quest to destroy the Ring, to overcome the evil
power of Sauron, and so restore the vitality of Middle-earth. Despite
its imagined pre-historical setting, the story, as in the Valentinian
system, is highly eschatological in orientation. Unlike the other
myths considered, both have a strong sense of development toward an
anticipated goal of global proportions and ultimate consequence. In
Tolkien's case, as we see with Sam's homecoming, the goal involves an
affirmation of the cosmos (re-greening of the Shire) and of embodied
human life (the evening meal with Rose and little Elanor). Though
heavily indebted to North European pagan myths, "*The Lord of the*

Rings," Tolkien wrote in 1953, "is of course a fundamentally religious and Catholic work; unconsciously so at first, but consciously in the revision.... The religious element is absorbed into the story and the symbolism."[21] In its origins as well as its historical development, Christianity itself is a creative synthesis of heterodox Jewish and pagan (Greek and more broadly Hellenistic) elements where, conversely relative to Tolkien's work, it is the pagan elements that get "absorbed into the story and the symbolism." It is arguably the manner in which orthodox (in contrast to Gnostic) Christian core symbols—notably that of the Incarnation—and values subtly pervade an otherwise pagan tale that determines the distinctive inflection of the monomyth in *The Lord of the Rings*.[22]

We will consider the particular relevance to the emergence of the Planetary Era of the symbol of the Incarnation, as well as that of the Trinity, in chapter 4.

3

FROM MYTH TO HISTORY

THE West does not have a monopoly on triadic schemes or triphasic patterns. There is a famous Zen saying regarding the path of enlightenment: "Before a person studies Zen, mountains are mountains and rivers are rivers [identity]; after a first glimpse into the truth of Zen, mountains are no longer mountains and rivers are not rivers [difference]; after enlightenment, mountains are once again mountains and rivers once again rivers [new identity]." Ken Wilber has shown how major Eastern philosophical and religious systems manifest variants of the triphasic developmental pattern, corresponding to the perennial distinction between body, mind, and spirit. There is also the related Vedantic distinction between the three fundamental states of consciousness of waking, dreaming, and dreamless sleep. These states correspond to the three broad realms of gross, subtle, and causal consciousness/reality. (A highly elaborated version of this is the Mahayana Buddhist *Trikaya* or doctrine of the three "bodies" of the Buddha: the "appearance" body, or *nirmanakaya;* the "enjoyment" body, or *sambhogakaya;* and the "truth" body, or *dharmakaya.*[23]) Unlike the Zen example, however, the third state or truth body does not necessarily include or integrate the first two. Where there is such inclusion, one can speak of the "fourth state" or *turiya* as the integral realization "of the primordial mind, the nondual mind that is ever-present in all sentient beings."[24]

In contrast with most Eastern (and most pre-modern) variants on the triphasic character of process or development,* the Hegelian-inspired triad is not limited to the religious or psycho-spiritual domains—though it is relevant to these as well. Central to the Hegelian outlook is the sense of evolutionary and *historical* unfolding. In fact, one could say that, for Hegel and those who share his intuition, it is only in the context of a philosophy of history, or more broadly from the point of view of a cosmologically and historically embodied evolution of consciousness, that the religious or psycho-spiritual domains reveal their true nature, their *Grund,* or raison d'être. This is equally true for the majority of Campbell's examples of the monomyth, which, even in his masterful hands, remain largely confined to the mythic-imaginal realm. While the idea of the monomyth opens a window to different times and cultures, Campbell's approach tends to do so from the ab-original perspective of the archetypal realm, the pre- or a-historical time of the beginning (*in illo tempore,* "in that time when..."), where all our journeys trace the identical circular pattern of the eternal return[25] of the seasons and the stars overhead. As Campbell himself puts it in his last work, *The Historical Atlas of World Mythology:*

> Mythology is not history, and history is not myth. Mythologies are products of the human imagination, which in turn is grounded in the energies that animate the organs of the human body, and since these have not significantly changed since the appearance of Cro-Magnon Man, c. 40,000 B.C., there can be recognized in the mythologies and associated ceremonials of all mankind certain themes and images that are constant [preeminently among which, the hero's journey or monomyth]. These are the "archetypes" of Jung, the elementary ideas of Bastian. They are not denotative of historical events, but symbolic powers, potentials, and requirements of the psyche, or to use an earlier term, the soul.[26]

---

* Aurobindo might perhaps be considered the great exception, though he wrote in English and was educated at Cambridge during the heyday of neo-Hegelianism, making his work a kind of dialectical synthesis (East/West, Human/Divine).

While appreciating the value of this perspective—it is indeed a
failure of imagination to reduce myth to (a literalized) history—I find
the way Campbell advocates an absolute split between history and
myth ultimately misguided. His position is admittedly complex (and
perhaps inconsistent), despite his ultimately dualistic view of the relation of myth to history and his elevation of the former over the latter.
Campbell sometimes seems to adopt the Jungian/Romantic Idealist
view of the evolution of consciousness as involving the differentiation of the ego out of an archetypal ground. And though he values
individuality and pluralism, tolerance and freedom (all emblematic of
modernity) and has a fine sense for the critical and planetary character
of late modernity (in his last and most influential work, *The Power of
Myth*, he becomes a kind of prophet of the nascent Gaian Age), on the
whole he tends more consistently toward a denigration of time and
genuine development. This denigration is reflected, for instance, in his
belief in the superiority of the Hindu theophany over the Abrahamic.

To my mind, it is essential, in this most challenging phase of the
Planetary Era, that we gain as much insight as possible into the specific trajectory that leads to the present. In the spirit of Hegel, as well
as in that of Joachim of Fiore (and of such diverse thinkers as Marx,
Comte, Gebser, and Jung), I believe that the fundamental triphasic
pattern behind or beneath the monomyth is evident not only in the
myriad examples of the hero's journey that Campbell has illuminated,
but also as the deep structure of Western—and, for better or worse,
world or planetary—history. The significant difference here is the following: whereas the monomyth *as myth* begins in the world of "common day" and ventures out into "wonder," as *history* (and as individual psychological development since modern times) it is the reverse:
it begins (first phase or moment) in wonder and ventures out (second
phase or moment) into the general disenchantment of common day.
This difference relative to Campbell's formulation is a reflection of
the rise of historical (and disenchanted), as distinct from mythical,

consciousness in the modern period, the period of the second phase or zenith of the larger arc.

An understanding of this triphasic pattern is also (as we shall see in chapter 12) critical to the search for a principle that might guide the expression of a genuinely planetary wisdom. It will be to the first and third moments of this pattern that we must turn for the clearest indications of the precursors to this wisdom. These moments coincide with the most concentrated expressions of a particular countercultural, and yet ultimately also planetary, trajectory of the evolution of consciousness.

The division of world history into three main epochs is, of course, widely familiar with the sequence: ancient, medieval, and modern. The word *modern* comes from the Latin *modo*, meaning "in a certain manner, just now." Thus *modern* originally meant "the present" or "recent times." It was certain Renaissance scholars who, in their enthusiasm for their wondrous present, considered the period from the fall of Rome to the "rebirth" of the spirit of the ancients to be a kind of purgatorial, "dark," or simply "middle" age. As we shall see in chapter 5, the birth of the modern period can indeed be seen as constituting a third (and simultaneously new first) phase in the spiral of the evolution of consciousness. From the eighteenth-century Enlightenment into our own times, however, *modern* is increasingly colored by the cluster of associations—secularism, materialism, disenchantment; Barfield's "idolatry," in other words—characteristic of the second phase of the larger arc.

Two more recent variants include: premodern, modern, postmodern, and pre-historical, historical, post-historical. Most writers who use these two sequences, however, though critical of the values or worldviews associated with the second term ("modern" or "historical"), reject the idea of a truly constructive, synthetic, or "final" (in Barfield's sense of "final participation") third phase. Although each of these threefold divisions of world history tells us something

significant, neither of them explicitly manifests the logic of the deeper
triphasic structure. A notable exception here is to found in the work
of David Griffin, John Cobb, and others inspired by the philosophy
of Alfred North Whitehead. It was Griffin who proposed the idea
of a "constructive postmodernism," which, as Cobb admits, "is, of
course, just one of several forms of postmodernism."

> The term *constructive* is used to contrast with *deconstructive* to
> emphasize that constructive postmodernism is proposing a posi-
> tive alternative to the modern world. This does not mean that it
> opposes the work of deconstructing many features of modernity.
> The point is that critique and rejection should be accompanied by
> proposals for reconstruction.[27]

These proposals, inspired as they are by Whitehead's "philosophy of
organism," fall squarely (or perhaps more appropriately, roundly) in
the spirit of a genuine third phase.

A fourth possibility for a threefold division, not just of world
history, but of the human journey on planet Earth, a division that
more clearly manifests the more fundamental pattern, might be:
anthropogenesis, diaspora, convergence. The first moment—the
birth of our species, *Homo sapiens sapiens*, perhaps some two hundred thousand years ago—was immediately followed by a great
diaspora in which, as Morin says, humans "became strangers to one
another through distance, language, rites, beliefs, and mores."[28] A
mere five hundred years ago, the scattered members of the human
family began a grand convergence with the beginnings of the
Planetary Era (this convergence, of course, has also involved the
extermination of much of human diversity, and increasingly now,
overall biodiversity, through domination by the Western colonial
cultures who initiated contact).

Karl Jaspers, in his *Origin and Goal of History*, also discerns three
great ages: prehistory (from the mystery of anthropogenesis to the

first city-states), history, and world history proper (corresponding to what I call *the Planetary Era*). "What we call history," he writes,

> and what in the old sense is now at an end, was an interim period of five thousand years' duration between the settlement of the globe, which went on throughout the hundreds of thousands of years of prehistory, and the beginning of world history properly so called, in our own time [that is, c. 1500 B.C.E.]... Our subsequent brief history...was like a meeting and gathering of men for the action of world history; it was the spiritual and technical acquisition of the equipment necessary for the journey. We are just setting out.[29]

While the "technical acquisition" would only arrive with the modern period (especially science and technology), the spiritual initially broke onto the world scene some two millennia earlier during the Axial Period (which, as we have seen, ranges from 800 to 200 B.C.E. in its first and most striking phase). With the near simultaneous emergence in the sixth century B.C.E. of the first Greek philosophers (from Thales and Pythagoras to Aristotle), the Buddha, Mahavira, Confucius, and Lao Tzu, the great Jewish prophets (Second Isaiah, Ezekiel, Jeremiah), and possibly Zoroaster, this period "gave birth to everything that, since then, man has been able to be, the point most overwhelmingly fruitful in fashioning humanity." It is during this period "that we meet with the most deepcut dividing line in history. Man, as we know him today, came into being."[30]

Echoing Jaspers' crucial insight into the deep structure of world history, and in a clear evocation of the fundamental pattern, W. H. and J. R. McNeill write that "human history is an evolution from simple sameness [identity] to diversity [difference] toward complex sameness [new identity]."

> Our remotest ancestors lived in simple, small groups, spoke only a few languages, and pursued a narrow range of survival strategies in East Africa.... Later..., people developed more social complexity,

reflected in a broad range of political forms.... The trend was toward more cultural differentiation, toward heterogeneity, toward islands of complexity in a sea of near uniformity.... At some point (I would guess between 1000 and 1 B.C.E. [that is, centered in the Axial Period]) the trend reversed. Interactive webs reduced cultural diversity.... As the webs grew and fused, complexity became the rule—the new uniformity.... This process is not complete nor is it likely ever to become so. Nonetheless, it is a striking trend of the last two or three millennia and probably has far to go yet before it reaches its limit or is somehow reversed.[31]

Earlier in the same book, the authors also recognize the large-scale shift from the second to the third moments in a manner consistent with Jaspers' understanding of the transition from "history" to "world history" or the Planetary Era. "In the three and a half centuries after 1450 [C.E.], the peoples of the Earth increasingly formed a single community. From this point forward, it makes less and less sense to treat different regions of the Earth separately."[32]

In his *World Religions and Social Evolution of the Old World Oikumene*, Andrey Korotayev points to the Axial Period as the transition in dominance from predominantly objective factors (for example: climate, availability of food) to more subjective ones (civilization in general, and religious beliefs, in particular). Because the rate of macroevolution of archaic and pre-Axial societies was slow and generally unnoticed (absence of literacy is obviously relevant here), "it missed their consciousness and was not an object of their concerns and interests."[33] During and following the Axial period, the first world religions and philosophies became the vectors for "the formation and diffusion of ideas implying that society changes substantially in time, that *ought to be* is substantially different from *what is*, that a more just social system is possible and that it could be attained through conscious human efforts."[34] What we have, in other words, is a differentiation and strengthening of human agency and of ideal or psycho-spiritual factors as determinants of social change. From this point on, world history and social macroevolution cannot be understood apart from a consideration of the evolution of consciousness.

William Irwin Thompson has proposed a particularly insightful recasting of world history and social macroevolution in terms of the evolution of consciousness. In his more recent writing,[35] he signals six great "transformations" embodied in seven distinct "cultural ecologies." The transformations run from hominization, through symbolization, agriculturalization and civilization, to industrialization and finally planetization. The seven cultural ecologies (which I group under the six transformations) are as follows:

1. Hominization:
    i. Sylvan (prehominid evolution of *Ramapithecus*)
    ii. Savannan/Lacustrean/Coastal (from *Australopithecus* to *Homo erectus*)
2. Symbolization:
    iii. Glacial (from archaic *Homo sapiens* to modern *Homo sapiens*)
3. Agriculturalization:
    iv. Riverine (ancient civilizations)
4. Civilization
    v. Transcontinental (classic civilizations)
5. Industrialization:
    vi. Oceanic (modern industrial nation-state societies)
6. Planetization:
    vii. Biospheric (planetary noetic polities)

In his many books and articles, Thompson illuminates the correlations between these transformations and cultural ecologies with various sets of categories that amplify his understanding of the evolution of consciousness. These sets include forms of social organization, governance, communication (e.g., oral, alphabetic, electronic), and of overall mentality. The latter are especially evident in the aesthetic (and, from the agricultural transformation onward, literary) and mathematical expressions typical of each cultural ecology. There are five of these overall mentalities:

1. Arithmetic (which presumably emerges during the transformation of Symbolization)
2. Geometrical (emerges with Civilization)
3. Algebraic (emerges during the late Classical)
4. Galilean Dynamical (bridging the late Classical with the Industrial)
5. Complex Dynamical (mid- to post-industrial)

I must refer interested readers to Thompson's own rich writings for a detailed discussion of these mentalities. Before moving on, however, I would make a couple of observations. Thompson himself points to the overlap of his five mentalities with Gebser's five "structures" of consciousness (archaic, magical, mythical, mental, and integral), the names of which give a general sense of how Thompson understands the five mentalities. Gebser, for his part, frames the five structures within an overall triphasic movement of collective consciousness from the pre-perspectival (archaic, magical, and mythic), through the perspectival (mental), to the aperspectival (integral). While Thompson does not highlight this overall triphasic movement, it is implicit in his thinking. What Thompson does do, however—and in a way, which, to my mind at least, represents a significant advance over Gebser—is explicitly link the third, integral-aperspectival phase (Thompson's Complex Dynamical Mentality) with the notion of the *Planetary*.

A question naturally arises at this point: If the notion of the Planetary is an expression of the (if not absolute, then at least proximate) telos or goal of the evolution of consciousness, can we see this goal prefigured in the first of the three phases? Thompson shows how the second phase (the mental-perspectival, especially as embodied in modern science, technology, and sociopolitical organization), paves the way (literally) for the emergence of beginning planetization in the form of the global techno-industrial complex.* But how can we understand, at its deeper levels, this beginning planetization relative to the pre-perspectival phase (and its mythic structure in particular) that preceded it?

In my previous work,[36] critically integrating Hegel's philosophy of history with Jung's psychology of individuation, I showed how the more fundamental triphasic pattern, which for Jung has to do with the differentiation of the ego or separate-self sense out of the archetypal unconscious and its ongoing approximation to the nature of the true Self is, for Hegel, apparent in the overall movement of world history in the shift in cultural dominance from the "East," through the Greco-Roman empires (where one witnesses the birth of the autonomous individual, or at least its ideal as applied to the elite class of male citizens of the polis), to "Europe" or the Western/Christian nations. It is here that the principle of freedom is explicitly adopted as the guiding ideal—however distorted and even mendacious in practice—of the national and international agendas. While both Hegel and Jung see prefigured in the Christian revelation the symbol of Spirit or the Self as the motive force and ideal

---

* Thompson also argues that the birth of modern science needs to be understood in relation to the creative impulse of the Western esoteric tradition, the spirit of which stands counter to the worldview(s) that come to dominate the modern period. Anticipating developments we shall track in the following chapters, Thompson notes: "Although William Blake chose to speak of 'single vision and Newton's sleep,' Newton and Boyle were as much Hermetic mystics as Blake, and the play of a cosmic and visionary mysticism has always been present in the cultural history of science. Newton had his box of Cabalistic computations on the temple in Jerusalem, and Descartes had his prophetic dreams.... It was the British technologists, the French positivists, and the German academics that all came together in the nineteenth century to give us the European orthodoxy in which one line of descent was considered legitimate and the other, the illegitimate offspring, one did not discuss in good company. Ironically, it was the physicists, the highest of the high priests of matter and scientific materialism, who enabled us to break out of the grip of the positivists, for with the indeterminacy of Heisenberg and the relativity of Einstein, the imaginative quality of the descriptions of science began to be undeniable." (Thompson, 1991, 13–14).

goal of world history, they also both agree that, with the advent of Christianity, this Spirit was realized in principle only. "The next point," as Hegel says, "is the development of this principle; the whole sequel of History is the history of its development."[37]

4

THE GREAT CODE

THE birth of the Planetary Era was the unforeseen—though, in retrospect, inevitable—culmination of what McNeill has called *the
rise of the West*.[38] Despite the diversity of European ethnicities and
then of nations, the whole of Europe and eventually the Americas
was covered by the "sacred canopy"[39] of the Christian mythos.
Michael Grosso[40] has shown to what extent the cultural history of
the West—from the spiritual Franciscans of the High Middle Ages,
through the Renaissance utopians, Enlightenment freethinkers,
and the American founding fathers, to Marxists, Nazis, New Agers,
and Futurists—has been shaped by what he calls the "millennium
myth," at the heart of which is a belief in the possibility of global or
planetary regeneration, the anticipation of "a new Heaven and a new
Earth." Similarly, Eugene Weber, in *Apocalypses* (1999), shows how
expectation of the end times runs as a continuous thread, or rather a
distinct and in some ways constitutive weave, throughout the fabric
of Western culture, both popular and "high." A list of famous figures for whom the apocalyptic or millennial theme was a significant,
and sometimes the most central, element of their worldviews would
include: Savonarola, Columbus, Tycho Brahe, Newton, Cromwell,
John Wycliffe, Jan Huss, Joseph Priestly, Joseph Fourier, Robert
Owen, Saint-Simon, Tsar Alexander (whose belief that Napoleon
was the Antichrist stimulated the idea for the League of Nations, the

precursor to the United Nations), and dozens more I leave to readers
of Weber's book to discover.

To understand the full import of this mythologeme of apocalypse
or the millennium, however, we must consider the larger Christian
symbolic matrix—including an esoteric reading of the symbol of
God as Trinity, and especially of the central symbol of Incarnation.
Properly interpreted, these symbols can be seen to have provided the
"Great Code"* for the eventual birth and ongoing transformation of
the Planetary Era. In doing so, we can also begin to see how this birth
and transformation might be understood as, to adopt Thompson's
phrase, "the planetization of the esoteric"[41]†

It is well-known the degree to which Hegel structured his system
according to a dialectical—and mystical—reading of the Trinity.[42]
Much of the early Christian theological imagination was devoted
to grasping the sense in which the one almighty God could, without diminishment, simultaneously manifest as Creator and Father in
Heaven, on the one hand, and on the other as Son and Redeemer—a
Son, moreover, who was a fully human being. While the doctrine of
the Trinity, with its distinction between the one substance and the
three persons (and the Western Christian belief in the co-generation
of the Spirit from Father and Son),‡ goes some way toward preserving
the essential monotheism, it does not, by itself, yield the central mes-
sage of the Great Code. For this we must look to the deep structure of

---

* "The Old & New Testaments," Blake inscribed as part of the commentary surrounding his engraving of the classical Laocoön sculpture, "are the Great Code of
Art."

† This is not the meaning of the phrase as Thompson uses it, though it is, I believe,
consistent with core elements of his thinking. As he uses the phrase, "planetization
of the esoteric" refers to the migration and incorporation, through the emergence
of planetary culture, of Eastern ideas (Buddhist and Taoist) into the technological
West. In a sense, what I am suggesting here is that we can interpret the phrase in a
reverse manner, where planetization is seen as a working out of what is already present, implicitly, as the esoteric core of a tradition (in this case, the biblical).

‡ This refers to the so-called *filioque* theological formula of the Catholic church.

the biblical worldview as a whole, with its three acts of Creation, Fall, and Redemption. Here we not only have a clear evocation of the fundamental pattern of identity, difference, and new identity, but also the pattern's expression as the generative principle of history (both cosmic and human) itself. At the center of the New Testament portion of the biblical drama is the mystery of Incarnation: the birth, passion, and resurrection (another expression of the three phases) of Jesus Christ, the eternal Logos become "flesh." This religious mystery was taken by Hegel to be a symbolic expression of the essential nature of Absolute Spirit or the evolutionary Whole—visible equally in nature and in history—as it was for Jung in the psyche's striving for wholeness or the actualization of the Self as *complexio oppositorum*, a "weaving together of opposites" (one of Jung's favorite formulas for the nature of the Self).

The paradoxical notion of the Incarnation was initially a stumbling block to many pagans and Jews, for most of whom there existed an absolute distinction, an ontological divide, between the realms of the (incarnate) human and the divine. We see this, for instance, in the Platonic contrast between the eternal realm of the Ideas (identified with the realm of the fixed stars) and that of becoming (the flux of the four elements, or everything below the Moon). This contrast, as we have seen, was most pronounced with the Gnostics. Despite its counterintuitiveness, there was something in the paradoxical quality of the new religion that offered a special appeal. Commenting on this appeal, Angus writes:

> In an eminently practical age when incarnate examples of human perfection were sought as guides to conduct and inspirations to higher endeavour, it is obvious that Christians had an immense superiority of appeal in the character, life, and ideals of one who had actually lived within recent times and died a martyr death, "crucified under Pontius Pilate." When an earnest Stoic could despairingly ask of his ideal Wise Man "Where is he to be found

whom we have sought so many ages?" the Christians could point
to one who was not only Lord in the cult, but the Elder Brother in
a numerous divine family [Romans 8:34]. The difference in point
of appeal and relativity to everyday life stands out in stark contrast
in the statements of two contemporaries, that of Plutarch, "This
[Wise Man] is nowhere on Earth nor has he ever been," and that
of the Prologue to the Fourth Gospel, "The Logos became flesh."⁴³

Christianity, he later comments, "early took on the character of a *complexus oppositorum*, which it retains" to this day.⁴⁴ *

It would appear that it is precisely the paradoxical, or complex,
character of the central symbol of the Incarnation that allowed the
Christian inflection of the biblical drama to become the Great Code
or generative monomyth for the emergence of the Planetary Era. The
argument here is itself complex and obviously dependent upon the
plausibility of the kind of (neo-)Hegelian and Jungian hermeneutic I am recommending.† An essential insight that this hermeneutic
reveals involves understanding how the modern, secular, and scientific worldview that followed upon the biblical, and that in many
ways represents its negation, is somehow an inevitable expression of
the incarnational descent into matter at the heart of the Christian
mythos. An early indication of how the idea of the Incarnation

---

* Here is Angus on the main elements of this *complexus:* "In this evolution from
primitive to universal form the Christian movement suffered all the agonizing of
the clarifying conflict between the literalists and the enlightened, between the
Judaizers and the Hellenizers, between the puritan and ascetic interpreters and the
humanistic and aesthetic,...between the institutionalist and his rigid adherence to
history and the mystic and philosopher who finds in the events of time and place
mere expressions of the eternal ideas" (Angus, 101).

† I say "neo" to distinguish my reading of Hegel from the British and American
Neo-Hegelians (esp. Bradley, McTaggart, and Royce) of the late nineteenth and
early twentieth centuries, whose positions did not benefit from the postmodern turn in philosophy and science. My own reading of Hegel has been shaped
by a consideration of Hegelian concepts (such as the dialectic, wholeness, the
Absolute, and the evolution of consciousness) in sustained dialogue with Jung's
depth psychology and Edgar Morin's views of complexity (see Kelly, 1993a; and
Kelly, 1988).

involves a creative self-negation is found in the Good Friday hymn of the early Lutheran, Johann Rist:

> O great distress, God himself lies dead,
> He died upon the cross,
> In this he won the kingdom of heaven
> for love of us.

Some three centuries later, Hegel would seize upon the phrase *God is dead* as an expression of the Absolute's creative self-negation, the dialectical movement whereby the wholeness symbolized by the Incarnation manifests its immanent, if initially paradoxical, actuality. The phrase is of course most (in)famously taken up by Nietzsche, who, in contrast to Hegel, wanted to do away with Christianity and all forms of transcendence altogether.

The rendering finite of the Infinite (the descent) has as its logical counterpart the exaltation of the finite. Jung has pointed to the shift in orientation from the medieval vertical (epitomized by the gothic cathedrals) to the modern horizontal (voyages of discovery, rise of humanism, study of nature) as expressive of the exaltation of the finite, with the intersection of both axes (forming a cross) symbolizing the continuing incarnation of the Christ principle or the actualization of the Self. [45] Hegel, for his part, understands the modern Western principle of freedom, conceived as the universal birthright of every human being, as having first entered Western consciousness through the New Testament teachings. Following the centuries of medieval cultivation of conscience, the principle bears secular fruit with Luther's defiance of the Papacy and the consequent political reorganization of Europe (I shall return to Luther in the following chapter).*

---

* Conforming Hegel's claim, and combining material with ideal factors in his analysis, Korotayev (2004) has presented extensive empirical evidence linking Christianity's early promotion of strict monogamy, its erosion of unilineal descent family organization, and its doctrinal stress on the practice of nonviolence with the rise in Europe of communal democracy.

The centerpiece of the modern worldview, however, is natural science. "Since the end of the Middle Ages," writes Jaspers, "the West has produced in Europe modern science and with it, after the eighteenth century, the age of technology—the first entirely new development in the spiritual or material sphere since the Axial Period."[48] The popular account of the origins of modern science is that, following Copernicus's stroke of genius in proposing a Sun-centered model of the universe, a quick succession of intellectual luminaries, from Galileo and Kepler to Newton, finally freed us from the limitations of medieval superstition and revealed the true nature of the cosmos. In fact, however, historians of science have revealed the extent to which many elements of the modern view of the cosmos were either explicitly anticipated or well prepared by Greek, Latin, and Islamic scholars and visionaries over the two millennia preceding Copernicus.[47] It was the Greek, Aristarchus of Samos (c. 270 B.C.E.), for instance, who seems to have been the first to present an argument that the Sun, and not the Earth, was the center of the cosmos. Though the more sophisticated model of Copernicus did not yield any more precise observational predictions of the movement of the planets as compared with the standard geocentric model of Ptolemy, it conformed more closely to the ancient Greek (and specifically Pythagorean) ideal of uniform circular motion, as well as to the Hermetic (and generally Platonic) high estimation of the Sun.

Just as the rise of Christianity must be seen in the context of the syncretistic environment of late antiquity (the creative encounter between Jewish, Greek, and Roman culture), so the rise of modern science cannot be understood apart from the fertile dialogue among Christian and Muslim (and to a lesser extent, Indian and Chinese) cultures during the Middle Ages.[48] Still, the fact remains that the scientific revolution did not occur in ancient Greece (or in India, China, or Baghdad), but in Western Europe. Many complex arguments have been proposed to account for this fact. In his *The Rise of Early Modern Science: Islam, China and the West*, sociologist Toby Huff[49] points to the unique character of the new European universities that institutionalized rationality in law and education. In the Muslim world, by contrast—and this despite the early lead over Europe in terms of mathematics, optics, observational astronomy, and access to Aristotle—no such institutionalization was possible because of the absolute and sequestered character of Islamic law.

Coming from the perspective of the history of hermeneutics, or the philosophy of interpretation of texts, Peter Harrison[50] argues that the view of nature presupposed by modern science was shaped by the shift in the way the Bible was interpreted following the Protestant Reformation. The new emphasis on literal interpretation of scripture spelled the end of the more traditional symbolic and allegorical readings. This emphasis was transferred from the Holy Book to the Book of Nature, which could now become a collection of literal and fully quantifiable "things" or objects.

The most suggestive account of the Christian origins of modern science, however, at least in terms of the idea of the generative power of the Bible as Great Code, comes from an essay by the twentieth century Russian neo-Hegelian, Alexandre Kojève. "Whether we like it or not," he writes, "the instigators of modern science were neither pagans, nor atheists, nor as a rule even anti-Catholics."

> It was because they fought, as Christians, against the ancient and pagan science that the likes of Galileo were able to elaborate their new science which, because it is our own, is still "modern.".…[51]
>
> In effect, what is the Incarnation if not the possibility that the eternal God is actually present in the temporal world in which we ourselves live, without thereby diminishing his absolute perfection? If, however, being present in the perceptible world does not damage this perfection, it is because this world is (or was, or will be) itself perfect, at least to a certain extent (an extent, moreover, which nothing prohibits from being determined with precision). If, as believing Christians affirm, a terrestrial (human) body can be "at the same time" the body of God and thus divine, and if, as the

Greeks thought, divine (celestial) bodies correctly reflect the eternal relations among mathematical entities, there is no longer anything to prevent us from searching for these relations here below as well as in the heavens. Now, it is precisely to such a search that more and more Christians passionately devote themselves from the sixteenth century onward...."[52]

In any case, it is Copernicus who eliminated from science all traces of pagan "docetism" [the heretical belief, found in many Gnostics, for instance, that Christ did not really suffer and die, that in fact he had no real physical body], by following the resurrected body of Christ in the Heavens with the whole terrestrial world where Jesus died, after having been born there [*en faisant suivre dans le Ciel le corps du Christ ressuscité par l'ensemble du monde terrestre ou Jesus est mort, après y être né*]. Now, whatever Heaven might be for believing Christians, for every scientist of the era it was a "mathematical" or "mathematizable" one. To project the Earth into such a heaven was thus equivalent to an invitation for these scientists to set upon, without delay, the immense (but by no means infinite) task of elaborating a (universal) mathematical physics. This indeed is what the Christian scientists proceeded to do. And since they did it in a world already largely Christianized, they could do so without seeming insane or risking scandal.[53]

What I take from such reflections is not the simplistic and rightfully contestable claim that Christianity, whether in its Catholic or Protestant forms, directly caused the Copernican Revolution and the rise of modern science. It is a question, rather, of the subtle but pervasive effect on the Western mind of a thousand years of exposure to the major Christian symbols (especially that of the Incarnation). This exposure seems to have prepared the imagination of the scientists (or *natural philosophers,* as they were then called) in such a way that previous insights and contributions, and those of the Greeks in particular, could be taken up and systematically articulated in a novel, and increasingly planetary, context.

With the above considerations in place, we more clearly see the underlying dialectical logic in the procession of dominant Western worldviews, which we can now list as follows:

I. IDENTITY: BIBLICAL/THEOLOGICAL	II. DIFFERENCE: MODERN SECULAR	III. NEW IDENTITY: PLANETARY
"Great Code": creation, fall, redemption; Incarnation: birth/passion/resurrection= descent of Logos	humanism; science; eventual disenchantment = exaltation of the finite	

The special characteristics of the third, planetary, worldview will become progressively more apparent as we proceed, in the following chapters, to consider in greater detail the birth and vicissitudes of the modern.

PART TWO

FROM ARC TO SPIRAL

"We shall not be what we have been, but we shall begin to be other."
—JOACHIM OF FIORE

"It is not difficult to see that ours is a birth time and a period of transition to a new era. Spirit has broken with the world it has hitherto inhabited and imagined, and is of a mind to submerge it in the past, and in the labor of its own transformation."
—G. W. F. HEGEL

"...twenty centuries of stony sleep
Were vexed to nightmare by a rocking cradle."
—W. B. YEATS

"To get ready for this new planetary culture, we climb and turn on the spiral and blink our eyes in wonder and disbelief as we see a history we missed in the settled cities of the plain where the universities lie."
—WILLIAM IRWIN THOMPSON

5

BIRTHING THE MODERN

MODERNITY, as we have seen, emerged as an organic expression of the evolution of Western culture. Much as the DNA and developmental unfolding of organisms reflect the deep structure and history of the species to which they belong, so the continuing evolution of Western, and now planetary, culture reflects the fundamental triphasic pattern to whose second phase modernity corresponds. As we shall see in the following chapters, the movement to the modern and toward the postmodern or planetary has proceeded by way of a fractal repetition of the fundamental pattern. A pattern is fractal when there exists a "self-similarity" between the properties of the overall or larger-scale shape and the parts or regions of which this shape is constituted. An example would be the shape of a mature tree relative to its branches and leaf-vein patterns, or the edge of a coastline as seen from space as compared to a much smaller part of the same edge viewed from closer up. In this case, one sees the fundamental pattern repeated once between the first and second phases (ancient to modern), and four times so far between the second and the third (modern to postmodern or planetary).

To begin with the first: We may take the early Christian community, as portrayed in Acts, as representative of the founding (ideal) identity of what was to become the dominant (and in many ways very far from this ideal) Western worldview—a community guided by what it took to be the spirit of the risen Christ, possessed by the ideal of unconditional love and the minimization of private property, and gifted with the power to heal and comfort those most in need. One can then trace the course of the fractal second moment in the community's progressive differentiation—which in this case can be read as an expression of continuing incarnation—into the medieval church as the dominant secular power of Christendom. In the first place, this involved what might be described as the eventual vaticanization of the original community: the establishment of an autonomous city-state within Rome, but whose power and influence reached to the far corners of the Western world; a rigid internal hierarchical structure; also, however, the investment of enormous amounts of energy into various cultural productions, most notably the great cathedrals, but also including numberless objets d'art, whose sensuous appeal stands in fitting counterpoise to the dominantly otherworldly orientation of the worldview.

It is during this period, at the midpoint of the High Middle Ages, that the Christian understanding of the Great Code undergoes a profound and consequential mutation. As a result of two visionary experiences while studying the Trinity, the Italian Cistercian monk, Joachim di Fiore (1135–1202), seems to have been the first explicitly to transpose the Trinitarian framework onto the overall structure of history. There are, according to Joachim, three great Ages, each named after the persons of the Trinity. The Age of the Father ended with the coming of the Christ. The age of the Son ended, and the Age of the Spirit began, in the time of Joachim himself. This New Age of the Eternal Gospel meant, among other things, that the rule of the Church hierarchy was to be replaced with that "of a monastic community of saints in the succession of St. Benedict, destined to cure, by an ultimate effort, a disintegrating world."[54] Some of the triads that Joachim associated with each of the three Ages include: childhood, youth, and maturity; law, grace, and greater grace;* starlight,

---

* A paradigmatic New Testament source here is Paul's Letters to the Romans, which highlights the shift from Law to Grace (corresponding to Joachim's second and third Ages).

moonlight, and daylight; fear, faith, and love. Anticipating two of the three ideals (*liberté, égalité, fraternité*) associated with the French Revolution, there is: bondage, *freedom*, and *friendship*.[55] This link with the much-later revolutionary triad is not fortuitous. While the radical groups inspired by Joachim—most notably the so-called Spiritual Franciscans—were condemned as heretics and either exterminated or forced underground, the spirit he released, or helped to channel—though neither foreseen nor intended by him—would eventually transform the face of the planet. "Joachim," writes Löwith, "like Luther after him,"

> could not foresee that his religious intention—that of desecularizing the church and restoring its spiritual fervor—would, in the hands of others, turn into its opposite: the secularization of the world which became increasingly worldly by the very fact that eschatological thinking about last things was introduced into penultimate matters, a fact which intensified the power of the secular drive toward a final solution of problems which cannot be solved by their own means and on their own level. And yet it was the attempt of Joachim and the influence of Joachim which opened the way to these future perversions; for Joachim's expectation of a new age of "plenitude" could have two opposite effects: it could strengthen the austerity of a spiritual life over against the worldliness of the church, and this was, of course, his intention; but it could also encourage the striving for new historical realizations, and this was the remote result of his prophecy of a new revelation.[56]

Despite Joachim's immediate and later profound, if subtle, influence on the evolution of Western thought and spirituality, the dominant trend in the centuries surrounding him (that is, from the ninth through the fifteenth centuries) is a deepening and honing of the intellect, what Jung would call the differentiation of the thinking function. We see this especially in the flowering of scholastic theology, culminating in Aquinas's fusing of Christian doctrine with Greek philosophy (especially the epistemologically and cosmologically oriented

Aristotle). This differentiation was instrumental in the eventual emergence of modern science, which, as we saw in the previous chapter, presupposed the creation of a stable community of inquiry that could match and eventually surpass the pagan academies. Medieval theologians not only made critical theoretical contributions to physics and astronomy (as, for instance, with Oresme's impetus theory and his proposal for the diurnal rotation of the Earth), but also to the modern scientific method in general (as we see with Roger Bacon's combination of mathematics and ingenious physical experiments). Moreover, as the theological tradition saw the emergence of what Tarnas calls "critical" scholasticism (especially William of Occam), essential foundations were laid for the eventual separation of reason and faith—a central tenet of the modern scientific worldview. [57]*

While modern science came to dominate the second phase of the larger arc in the evolution of the Western worldview—a phase that represents a clear differentiation, and even, from the time of the eighteenth-century Enlightenment, an outright dissociation from the faith-based orientation of medieval Christendom—modern Western consciousness clearly also constitutes a new identity (or a cluster of related new identities). This is apparent not only with the modern

---

* Bruno Barnhart summarizes the emergence of the "new autonomy of the 'natural'" as follows: "(1) in the new 'gospel liberty,' manifested in a general way by the new disciples of the vita *apostolica;* (2) in freedom from the sacred enclosure of monasticism, a system of rules and customs that had confined human life; (3) in a renewed valuation of the literal sense of the Scriptures; (4) in the autonomy of human reason— a liberation from confinement within the biblical world of thought, to generate a 'human' system of theology (this was expressed in the new *quaestiones*, in the free space if disputation that succeeded the traditional silent assent to authority); (5) in the freedom to preach the gospel where it was needed; (6) in a new freedom of the divine word itself from the sacred structures that had fettered its dynamism and muted it; (7) in a new movement of historical movement, from the static structures of the church and of traditional consciousness; (8) in a freedom from the sacred distinctions that had separated clergy from laity; (9) in freedom from sacred traditions of biblical interpretation and theological construction; (10) in a growing confidence in the autonomy of secondary causes; (11) in a new autonomy and differentiation of the secular political order, as exemplified by the desacralizing of empire and civil authority, as well as by the cessation of the tradition of vassalage" (Barnhart, 107).

scientific temperament, but in the two main movements that precede and accompany the Copernican Revolution from which modern science takes its formal origins: the Renaissance and the Reformation.

Though the name *Renaissance* is generally taken in the sense of a rebirth of pagan learning, the movement can equally be understood as indicating both a rebirth of European Christendom and as a new birth in its own right.[59][*] All new births, of course, draw many of their traits from their parents—in this case, the Christian theological tradition and various freestanding elements of ancient Roman and Greek culture (many of the latter mediated by Islamic scholars). What is new in the Renaissance is not so much the synthesis of pagan and Christian values and motifs. As we have seen, Christianity is synthetic, or syncretistic, from the start. Rather, it is the overall character of the Renaissance spirit, especially in its more esoteric inflections (I am thinking here of Ficino, Pico, Agrippa, Paracelsus, Bruno, Campanella, Dee, Boehme, and Fludd, among others), which one could describe as organicist, as embracing of an expanded and enchanted cosmos, an exalted view of human potential, and a utopian social vision. Though there are many and sometimes substantial differences among views, common to all, as Merchant notes, "was the premise that all parts of the cosmos were connected and interrelated in a living unity."

---

[*] Commenting on the nineteenth-century French historian Jules Michelet, Reeves and Gould (Reeves and Gould, 1987) note: "A striking feature of Michelet's perception of Joachim [of Fiore] is the way in which he links the medieval prophet with the burgeoning spirit of the Renaissance. Its roots, he held, lie back in the period of Joachim and the Eternal Evangel. This teaching...was the 'alpha' of the Renaissance, and it was with rare intuition that he linked both Savonarola and Michelangelo with Joachim's vision" (p. 69). More recently, Barnhart proposes "that the human and humanistic renaissance that is evident during this whole period is a historical expression of the Christ event. We seem to observe at this time a recurrence of the human rebirth that appears in the New Testament, but in a new Western context and with a distinctly worldward orientation. The twelfth-century rebirth is at once Christian and secular; this is one of the dualities, held in equilibrium, in which the new vitality of this time can be seen" (Barnhart, 104).

From the "affinity of nature" resulted the bonding together of all things through mutual attraction or love. All parts of nature were mutually interdependent and each reflected changes in the rest of the cosmos.[58]

"The whole world," wrote Della Porta, "is knit and bound within itself: for the world is a living creature, everywhere both male and female, and the parts of it do couple together…by reason of their mutual love."[59] Bruno, similarly, claimed that it is not reasonable "to believe that any part of the world is without soul, life, sensation, and organic structure."[60] One of Dee's aphorisms anticipates the holographic paradigm (of which I shall have more to say in chapter 11) of our own times. "Whatever exists by action," he wrote, "emits spherically upon the various parts of the universe rays which, in their own manner, fill the whole universe. Wherefore every place in the universe contains rays of all the things that have active existence."[61]

"In the opening years of that momentous seventeenth century," writes Frances Yates, "every kind of magic and occultism was rampant. The authorities were deeply alarmed."[62] Though minimized by several generations of later historians, the esoteric cosmological perspective of the Renaissance was crucial to the inspirations of Copernicus, Kepler, and even to Newton, despite the great secret he made of it. "For twenty-seven years," writes Michael White in *Isaac Newton: The Last Sorcerer*,

from 1669 until he left London in 1696, Newton pursued a vast collection of themes both scientific and alchemical, as well as subjects as seemingly diverse as biblical chronology, numerology, history and mythology. Newton seems to have led a double life in the 1670s—the scientific inquirer, moulder of a new approach to science…coexisted with occultist, the seeker of the ancient flame of wisdom and arcane knowledge. He juggled the responsibilities of his position in the academic world with his clandestine and totally heretical ideas, keeping hidden his unorthodox and socially unacceptable religious views.[63]

While Newton's alchemical studies show his continuity more with Hellenistic esotericism, and also with the nature-aligned dimension of the incarnational streams running through Renaissance esotericism, his preoccupation with apocalyptic number speculation and millennial prophecy is more expressive of the Jewish and history-aligned incarnational streams. In any case, both sides of his ruling esoteric passion were pursued in secret. Whatever individual psychological reasons might have been involved in Newton's double life, one has to bear in mind that the years in question overlap with the reactionary period of the Restoration. In the radicalized period surrounding Newton's birth (1642), by contrast, the occult traditions of the Renaissance were openly and widely pursued. According to historian Christopher Hill, "astrology, alchemy and natural magic contributed, together with biblical prophecy, to the radical outlook."[64] This was true not only of religious and political radicals, but of the members (such as Newton's older contemporary, Robert Boyle) of the so-called Invisible College, the precursor to the Royal Society. Radicals Gerrard Winstanley and John Webster, for instance, recommended that astrology be part of the official university curriculum.[65] "To know the secrets of nature," declared Winstanley,

> is to know the works of God.... And indeed if you would know spiritual things, it is to know how the spirit or power of wisdom and life, causing motion or growth, dwells within and governs both the several bodies of the Earth and planets in the heavens above: the several bodies of the Earth below, as grass, plants, fishes, beasts, birds and mankind.[66]

This period of radical thought and practice in England represents the point at which the last wave of the Renaissance intersects with the originally revolutionary spirit of the Reformation. Like the Renaissance, the Reformation gives expression to a profound re-rooting—in this case into the soil of primitive Christianity rather than that of classical Rome and Greece. At the same time, however, just as the re-rooting of the Renaissance helped give birth to a new, and in many ways unparalleled, vision of both human nature and the cosmos, so the Reformation sparked by Luther marked the emergence of a new principle of subjectivity based on the experience of immediately given religious feeling (*sola fide*, faith alone) and intensified conscience. Along with this radicalized subjectivity, and greatly amplified by both his German translation of the Bible and by the newly invented printing press, Luther's doctrine of the "priesthood of all believers" made for a new democratization of spirituality. Despite the tendency toward a greater reliance on a literal reading of the Bible as the standard of authority, these developments also signaled the emergence of a new ideal of freedom or autonomy relative to the more collectively and hierarchically oriented medieval mentality. Although still, as Luther put it, "the most dutiful servant of all, and subject to everyone," a Christian "is the most free lord of all and subject to none."[67]

Though initially confined to the sphere of religion or spirituality, this ideal was not without its consequences in the political sphere. It was generally in the interests of the German lords to side with Luther, thereby legitimizing their already developing desire to break the constraints associated with allegiance to both the Pope and the Holy Roman Emperor.* The Weberian hypothesis concerning the apparent links between the (Calvinist) Protestant work ethic and the spirit of capitalism are well-known, however contested or qualified by some.† Thus, two of the fundamental traits associated with the

---

* Luther, for his part, though he initially supported the cause of the peasants in their own revolts, eventually sided "Against the Robbing and Murdering Hordes of Peasants," as he put it in his virulent 1525 pamphlet. Some estimate that, by 1525, perhaps 100,000 of the rebellious peasants had been killed by the armies of the feudal lords.

† See the comprehensive review article and bibliography by John Munroe of the University of Toronto, www.economics.utoronto.ca/munro5/ProtCap2.htm (6/22/07). According to Munroe, Weber "denies that he believes that the spirit of capitalism could only have derived from the Reformation, and claims that he

modern worldview—the ideal of freedom and the disciplined pursuit of capital—are intimately tied to the Protestant Reformation, which transpired alongside the rise of modern science.

The first fractal repetition of the fundamental pattern, therefore—that is, the one between the first and second phases of the greater arc—can be summarized as follows.

First Cycle

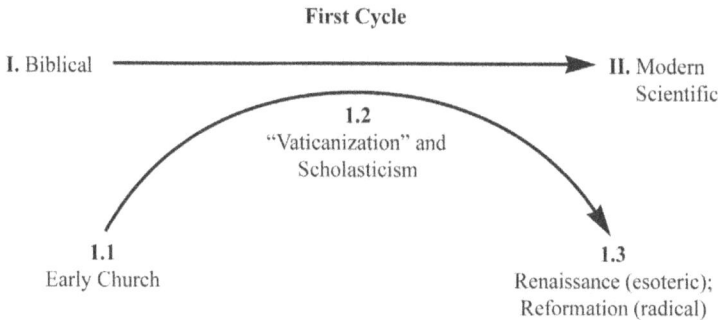

I. Biblical   II. Modern
Scientific

1.2
"Vaticanization" and
Scholasticism

1.1
Early Church

1.3
Renaissance (esoteric);
Reformation (radical)

While the modern impulse is often seen as running counter to the thousand years of Christendom that preceded it, we have seen that it is more coherently understood as a dialectical expression of the incarnational spirit of the Great Code with which Christendom was deeply informed. Many elements of what will later be associated with the sixties Counterculture, however, *are* present in the English radicals before the Restoration. We will consider this Counterculture and the New Paradigm that issued from it in more detail in chapters 7, 10, and 11. Here I would point to the general sense of social and cultural revolution, the ideals of communalism, free love, social justice, and the value of personal experience and altered states of consciousness.

---

only wishes 'to ascertain whether and to what extent religious forces have taken part in the qualitative formation and quantitative expansion of that spirit over the world.' Nevertheless, he often does suggest that it was Christian asceticism and Calvinism that provided the orientation that led to the development of such ideas as the 'necessity of proving one's faith in worldly activity,' 'the preaching of hard, continuous bodily or mental labor,' and 'rational conduct on the basis of the idea of the calling' that were to provide 'the fundamental elements of the spirit of modern capitalism.'"

Three centuries before the civil rights movement of the sixties, and more than a century before the American and French Revolutions, radical Puritans and other sectarians were calling for the abolition of private property and a kind of universal democracy. "Many called for a republic," writes Barzun:

> the vote for all; the abolition of rank and privilege; equality before the law; free trade and a better distribution of property...because these goals were justified out of Scripture, the substance of Puritan political thought has been eclipsed.... It is easier to credit John Locke than some obscure Anabaptist preacher for the thought that all men are born free and equal.[68]

"True freedom," declared Winstanley, "lies in the community in spirit and community in the earthly treasury, and this is Christ the true manchild spread abroad in the creation, restoring all things unto himself."[69] Hill notes that Winstanley, along with many others in early seventeenth century England, were inspired by Joachim of Fiore's millenarian vision of the dawning of the Third Age of the Spirit, in which

> "the Lord himself, who is the Eternal Gospel, doth manifest himself to rule in the flesh of sons and daughters." Their hearts will be returned to the Reason [the Logos; the *anima*, or *spiritus mundi*] which pervades the cosmos, to "that spiritual power that guides all men's reasoning in right order to a right end." Every man subject to Reason's law becomes a Son of God..."; his ruler is within, whether it be called conscience or love or Reason. This is Christ's second coming.[70]

Many who appealed to the "inner light" of the Christ within were branded as "enthusiasts." Like the ancient Gnostics, their reliance on immediate experience was a threat to established authority (both ecclesiastical and political). Unlike the Gnostics, however, most enthusiasts during the English revolutionary period were anything but anti-cosmic.

"God is in everyone and every living thing," said the Ranter, Bauthumley; "man and beast, fish and fowl, and every green thing, from the highest cedar to the ivy on the wall." He does not exist outside the creatures. God is in "this dog, this tobacco pipe, he is in me and I in him"; he is in "dog, cat, chair, stool."[71]

The reference to tobacco here is not fortuitous. It seems to have played a somewhat analogous role in these radical circles to marijuana in the 1960s. This analogy with modern drug taking, writes Hill,

> should enable us to understand that—in addition to the element of communal love-feast in such gatherings—the use of tobacco and alcohol was intended to heighten spiritual vision. Some years later the millenarian John Mason was excessively addicted to smoking, and "generally while he smoked he was in a kind of ecstasy."[72]

The English radicals and their programs for sweeping reform were effectively crushed or suppressed, beginning during Cromwell's Protectorate and more definitively following the restoration of the monarchy in 1660. There were some gains: most notably, the shift from an absolute to a constitutional (democratic) monarchy, as well as the strengthening of certain guilds (for example, those of the London transport workers). The religious impulse, however, was increasingly depoliticized. The Quakers pointed the way to a new emphasis on quietism and a more generally introverted spiritual perspective. With the transformation of the Invisible College into the Royal Society, the pursuit of modern science would proceed through the systematic elimination of the esoteric or occult elements (epitomized in the practices of alchemy and astrology) with which, as we have seen, it was originally and intimately bound. "The radicals of the English Revolution," Hill laments, "made the last attempt to see the universe as a whole [an organic and enchanted whole, I would add], science and society as one."[73] The defeat of the radical scientists "also meant the end of dreams of an all-embracing *Weltanschauung* accessible to ordinary people."[74] It was *not* the last attempt, however, as the following pages will show.

6

TRIUMPH, REVOLUTION, AND PROTEST

WHILE the early modern period of the Renaissance, the Reformation, and the Copernican Revolution represented a departure and differentiation from the more mythically embedded, self-enclosed medieval worldview, it was still generally Christian. The human being retained its nobility, even in its fallen state, by virtue of its divine and immortal soul. The cosmos (including the Earth) was still enchanted, even though "it" now seemed ready to reveal its deepest secrets to an increasingly sophisticated mathematical and technological gaze. This would all change, however, over the next few centuries as the modern Western mind sought not only to differentiate, but to dissociate itself from the matrix out of which it emerged. The dynamic is doubtless familiar to anyone who reflects on his or her own adolescence and the perfectly healthy, if often and inevitably deluded, belief that it was necessary or even possible to rebuild the world from scratch.

It was the explicit ambition of Descartes (1596–1650) to do away with all received opinion, and to achieve, through the example of mathematics and from the secure foundation of the experience of self-certainty, a total reconstruction of knowledge (even Descartes, however, believed that his whole project must fail were it not for the existence of God, the absolute substance). Though recognized as an arch-rationalist, Descartes was inspired in his quest for a new foundation by a series of dreams or visions that came to him on November 10, 1619. The account of his "enthusiasm," as Yates notes, reminds us of "the atmosphere of the Hermetic trance, of that sleep of the senses in which truth is revealed."[75] Descartes was well aware of the Renaissance Hermeticists as well as of the recently established Rosicrucians, though he was initially unsuccessful in finding out much about them. Descartes' friend and adviser, Marin Mersenne, spent the greater part of his writing career attacking Hermeticists and Rosicrucians (especially Fludd), condemning "the doctrine of the anima mundi, or at least the extravagant extension of this indulged in by the Renaissance naturalists who affirm that the world lives, breathes, even thinks."[76] May not, asks Yates, "the intensive Hermetic training of the imagination toward the world have prepared the way for Descartes to cross the inner frontier?"[77] This training involved "a new direction of the will toward the world, its marvels, and mysterious workings, a new longing and determination to understand those workings and to operate with them."[78] However much the Renaissance esotericists may have prepared the way for Descartes and the modern scientific project, however, Descartes broke definitely from them, rejecting their symbolic or analogical mode of reflection in favor of the analytic method he pioneered. He also abandoned their organic and enchanted worldview and in its place proposed the dualism of (self-reflexive) mind and (dead and mechanistic) matter with which he is most infamously associated.

Though similarly mechanistic, John Locke (1632–1704) stressed the role of sense experience in shaping an otherwise blank slate of a mind, and thus offered the possibility of restructuring the world on empirical grounds (though Locke, too, as much as he valued the power of the human mind, recognized the supra-rational nature of revealed truth). It was Newton (1642–1727), however, despite his obsession with matters esoteric and his heretical brand of Christianity (he was a closet Arian), who more than anyone else demonstrated the theoretical

---

* *Arianism* refers to the belief, founded by Arius, that Christ did not exist eternally like God the Father, but had a beginning in time (like the angels). It was condemned in 381 C.E. at the Council of Constantinople.

potential of the mechanistic paradigm. It would not be long before the successes of the "mechanical philosophy" would inspire belief in a more generalized view of *progress*.

Perhaps the most characteristic theme of the eighteenth-century Enlightenment, the idea of progress would be taken up and popularized by Voltaire (1694–1778) and the French philosophes. For the German Lessing (1729–1781), the spirit of Joachim is still clearly visible, though now thoroughly secularized, in the idea that the "education of the human race" (the title of his most influential nondramatic work) has been accomplished with the help of two "primers": the revelations of the Old and New Testaments. Lessing proclaims: "It will assuredly come! It is the time of a new eternal Gospel, which is promised us in the Primer of the New Testament itself!"

> Perhaps even some enthusiasts of the thirteenth and fourteenth centuries [he is referring to the Joachim-inspired "spiritual Franciscans"] had caught a glimpse of a beam of this new eternal Gospel, and only erred in that they predicted its outburst at so near to their own time.
>
> Perhaps their "Three Ages of the World" were not so empty a speculation after all, and assuredly they had no contemptible views when they taught that the New Covenant must become as much antiquated as the old has been....
>
> Only they were premature. Only they believed that they could make their contemporaries, who had scarcely outgrown their childhood, without enlightenment, without preparation, men worthy of their *Third Age*.[79]

For Voltaire, however—as for the Enlightenment philosophes in general—the biblical revelations were not so much primers as bad histories riddled with superstitions and barbarities. As for traditional religions, and organized Christianity in particular, his rallying call, "Crush the infamous thing!" (*Écraser l'infâme!*) says it all. Voltaire did not live, despite his long life, to see his wishes apparently fulfilled with

the official abolition of Christianity in May of 1794 by the new French
revolutionary government (the first day of the new year 1 had already
been proclaimed on September 22, 1792). Neither did Condorcet (1745–
1794), though an early leader of the Revolution and author of the most
famous treatise (published posthumously) on the Enlightenment ideal
of progress (*Sketch for a Historical Picture of the Progress of the Human
Mind*), who died in prison where he kept held for holding too mod-
erate views. It is a sign of the nobility of Condorcet's character and
convictions that he was able to compose his hymn to progress, which
culminates with the promise of the Revolution that he helped bring
about, while awaiting probable death by guillotine from the forces of
the Terror into which the Revolution had devolved. One must try to
imagine Condorcet in his cell, scratching out with his quill how we
must look with hope to

> the moment knowledge shall have arrived at a certain pitch in a
> great number of nations at once, the moment it shall have pene-
> trated the whole mass of a great people, whose language shall have
> become universal, and whose commercial intercourse shall embrace
> the whole extent of the globe. This union once having taken place
> in the whole enlightened class of men, this class will be considered
> as the friends of humankind, exerting themselves in concert to
> advance the improvement and happiness of the species.[80]

If the French Revolution at first seemed—to the sympatheti-
cally minded, at least—to signal a beginning realization of the
Enlightenment ideal of a free, rationally ordered, secular society, the
Terror, which immediately followed, revealed a fathomless shadow
behind, or beneath, Reason's light. Pre-revolutionary Europe, of
course, had already produced correctives and alternatives to the
rationalistic and mechanistic worldview—the so-called Cartesian–
Newtonian paradigm—which otherwise dominated the eighteenth
century: pietism in religious circles; sentimentalism in fiction and
poetry; a growing interest in the concepts of the picturesque, the

gothic, and the sublime (and with them a revaluation of the worth of the Middle Ages). As was the case with the radical sectarians that flourished during and immediately following the English civil wars, the events in France do, however, seem to have worked as a kind of catalyst for a more generalized shift or even mutation of consciousness, the two most significant manifestations of which, in terms of identifiable movements, were Romanticism and post-Kantian Idealism. Summarizing the nature of this mutation with respect to the periods that preceded it, Abrams writes that "faith in an apocalypse by revelation had been replaced by faith in an apocalypse by revolution, and this gave way to faith in an apocalypse by imagination [Romanticism] and cognition [Idealism]."[81] In both cases, "the mind of man confronts the old Heaven and Earth and possesses within itself the power...to transform them into a new Heaven and new Earth, by means of a total revolution of consciousness."[82]

Despite their distinctiveness in other respects, these two movements, which together announce the transition to the third moment in this second fractally repeated cycle within the larger arc, share a number of fundamental traits, most of which are also to be found in the corresponding third moment of the previous cycle (which, in this new cycle, now becomes the first moment) with the Renaissance esotericists and natural philosophers. Romanticism, writes Gusdorf, "returns to a previously dominant mode of knowing that does not limit its ambition to the deciphering of the superficial arrangement of phenomena, but instead strives to forge an alliance with the essence of cosmic reality. Rather than sticking to the dust of data...one must try to penetrate to the depths of being by means of a kind of knowledge that is also a wisdom and almost a religion."[83] For the Romantics and Idealists, instead of the machine, the *organism* and *life* become the root metaphors for the cosmos as a whole. This insight is particularly stressed in the tradition of *Naturphilosophie*, initiated by Schelling, but with a parallel tradition stemming from Goethe (the two figures in fact acknowledged a profound mutual affinity). For these *Naturphilosophen*, the whole Cosmos

> forms a nesting of living unities, each one included in the others.... Romantic biology, founded on the *analogy* of the universal organism, presupposes a global *harmony* which brings each part of the world into greater proximity with every other part. *Sympathies* and *antipathies* reveal the presence of each to all, of each in all, regardless of material distance.... While the mechanistic paradigm displays a series of mutually external chain links...the romantic paradigm evokes a liquid milieu where each site finds itself in contact with all others, a milieu through which waves of meaning, propagating in all directions, are capable of generating harmonics and positive or negative resonances.[84]

The case of Schelling in particular highlights the resonance between the Romantic/Idealist worldview and that of the Renaissance.* In contrast to the latter, however, this organicism also involves a new emphasis on the idea of *development*. As we have seen, it was Joachim who seems to have been the first to envision the dynamic or dialectical power of the Trinity at work in the progressive unfolding of the divine plan in and as history. The ancient and medieval tradition of alchemy, for its part, saw a kind of developmental potential in nature, which the adept sought to hasten with the dialectical procedures (especially separation and coagulation) of the "great work." With the Romantics and Idealists, it is as though these two traditions—the Joachimite and the alchemical—unite to produce the first explicit formulations of an evolutionary theory. "Within the tradition of *Naturphilosophie*," writes

---

* Schelling published a dialogue in 1802 entitled *Bruno, or, On the Natural and the Divine Principle of Things* (see Schelling 1984), named after the Renaissance visionary Giordano Bruno, who was burned at the stake in 1600 for various heresies, including his particular reading of Copernican heliocentric cosmology, which to him implied the view that there are an infinite number of solar systems; reincarnation; and his association with the heretical queen Elizabeth I (see Yates, 2002). For the relation between Romanticism/Idealism and esotericism more generally, see not only Abrams (pp. 141–163), but also Hanegraaff, 1998, pp. 384–420.

Robert Richards, "nature ceased to be the mere product of the Creator's designs [to the extent that a Creator was even recognized—Napoleon's physicist Laplace, it will be recalled, having "no need of that hypothesis"] but itself became producer—of itself.... Like a growing individual, it took on the form of a completely historical entity."[85]

Unlike what will come to be understood by the term, however, the Romantic and Idealist conceptions of evolution or development are not only organicist rather than mechanical, but also thoroughly *(re-)enchanted*. Nature, said Schelling, is slumbering Spirit. It is "the holy and ever-creating primal energy...which begets and actively produces all things from itself."[86] "Permeating everything," he declares, "there is but one and the same life, the same ontological power, the same ideal bond. In nature there is nothing purely corporeal, but everywhere the same soul symbolically transmuted into flesh."[87] The overall drive of the cosmos is toward the production of increasing complexity of organization as the vehicle for the eventual emergence of self-reflexive consciousness—again, however, not as an epiphenomenal or even merely emergent quality from an otherwise dead or inert matter,* for the order or organization of the cosmos is a manifestation of soul (the *anima mundi*), or Spirit (*Geist*). At the same time, the characteristically human expressions of this consciousness are not to be restricted to the mechanistic intellect (*Verstand*) of the Enlightenment, but now encompass a much vaster range of potentials, from the sub- or unconscious (a term that enters the language in this period) to the superconscious, from the creative imagination favored by the Romantics to the intellectual intuition or speculative Reason (*Vernunft*) of the Idealists.

---

* Where the notion of epiphenomenalism makes life and mind into wholly derivative by-products of matter, the more recent and increasingly popular notion of emergence involves the recognition that life and mind have distinct higher-order properties that, though dependent upon prior conditions of material organization, are not fully reducible to these prior conditions. For example: all organisms are self-organizing (self-producing and self-reproducing) in ways that none of their molecular components (carbon, hydrogen, nitrogen, oxygen, phosphorus, sulphur, and so on) are when considered in isolation.

The evolutionary dimension of this organicist and (re)enchanted worldview applies equally, and even more pertinently, to history and human consciousness as to the nonhuman cosmos, since it is only in and as the actualization of human consciousness that the full meaning of the evolutionary dimension reveals itself. In contrast to the dominant linear view of progress already articulated during the Enlightenment, the Romantic/Idealist view is more of the nature of a "spiral journey," as Abrams describes it, a journey whose milestones, as we read in his masterful summary below, were already laid down by the basic Christian mythos. Referring to Hegel and Schelling, the two greatest Idealist philosophers, Abrams writes:

> the timeless metaphysical system, as it evolves through time in the mode of history, has a clearly defined plot: the painful education through ever expanding knowledge of the conscious subject as it strives...to win its way back to a higher mode of the original unity with itself from which, by its primal act of consciousness, it has inescapably divided itself off. When described in metaphysical terms, the basic categories of this process...tend to be derived from the model of biological genesis, growth, and development; a process which was in turn interpreted as the sequential reconciliation of recurrent contraries....But in addition the historical process was often represented...in an allegorical or pictured form...as a circuitous journey back home. So represented, the protagonist is the collective mind or consciousness of men, and the story is that of its painful pilgrimage through difficulties...in quest of a goal which, unwittingly, is the place it had left behind...and which, when reachieved, turns out to be even better than it had been at the beginning. Thus redemption, even after it has been translocated to history and translated into the self-education of the general mind of mankind, continues to be represented in the central Christian trope of life as a pilgrimage and quest: the *Bildungsgeschichte* [history as education] of the Romantic philosophy of consciousness tends to be imagined in the story of a *Bildungsreise* [educational journey] whose end is its own beginning.[88]

The spiral path of this educational journey, as we have seen, can be understood as an expression of the dialectically generative potential of the fundamental pattern. The second fractal repetition of this fundamental pattern—this time occurring between the second and third phases of the greater arc—can be summarized as follows.

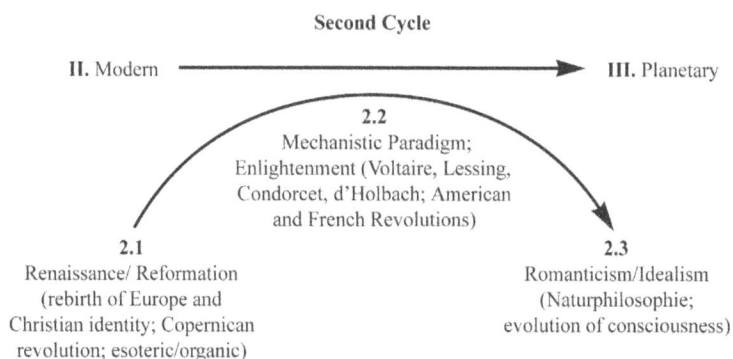

Second Cycle

II. Modern ⟶ III. Planetary

2.2
Mechanistic Paradigm;
Enlightenment (Voltaire, Lessing,
Condorcet, d'Holbach; American
and French Revolutions)

2.1
Renaissance/ Reformation
(rebirth of Europe and
Christian identity; Copernican
revolution; esoteric/organic)

2.3
Romanticism/Idealism
(Naturphilosophie;
evolution of consciousness)

Though the fundamental pattern is embedded as the "Great Code" of the Christian mythos (Phase I) and was even applied to history as a whole by Joachim some seven-hundred years earlier, it is with the Romantic poets and Idealist philosophers that the pattern first receives its mature and paradigmatic form. The period in which this takes place—roughly twenty or so years on either side of 1800—is staggering in the sheer number of creative figures who inhabited it, some of the more well known of which include: Blake, Wordsworth, Coleridge, Keats, Shelley, and Byron; Wollstonecraft and Godwin; Haydn, Mozart, Beethoven, Chopin, and Schubert; Goethe, Kant, Schiller, Herder, the Schlegel brothers, Fichte, Hegel, Schelling, Hölderlin, Heine, Novalis, Schleiermacher, and Schopenhauer; Jefferson, Franklin, and Madison. Given not only its extraordinary concentration of genius, but more especially because of the manner in which the fundamental pattern rose so clearly into the consciousness of its intellectual and spiritual leaders, the period in question can be considered as a kind of mini *axial age*.

In contrast to Jaspers's "Axial Period," however, this more concentrated mini-axial age spans just a few decades instead of more
than half a millennium,* and transpires on mostly one continent
instead of three (unless we include the Transcendentalist movement, the American counterpart to the European Romantic/Idealist
movement, in which case it would be two instead of one). A more
significant difference, however, is that, whereas the high point of the
great Axial Period falls two centuries short of including the birth
of Christianity, the mini-axial age of the Romantics/Idealists, as we
have seen, is both in organic continuity with, and represents a distinct (though equally organic) mutation relative to, the Trinitarian
and incarnational spirit of the Christian mythos.

Though distinctive and in certain respects even unique, we have
seen how this mythos emerged as a syncretistic union of two of the
great traditions of the first Axial Period—namely, the Jewish and the
Greek—which came into creative contact during the Hellenistic and
Roman periods of late antiquity.† The early modern period (in such
figures as Ficino and Pico through to Agrippa, Bruno, and Fludd)
sees a fresh infusion from these traditions—though now in their later,
more heterodox and esoteric (and now themselves also hybridized or
syncretistic), forms—with the discovery of the Hermetic corpus, on
the one hand, and the Kabbalah, on the other.

Despite a long history of sporadic contacts (from Alexander
the Great in northern India to Marco Polo at the court of Kublai
Khan), it is not until the beginning of the modern period, or

---

* Richard Tarnas (personal communication) has pointed out to me that the forty-
year period 590 to 550 B.C.E.—arguably the high-point of the Axial Period—is perhaps equally concentrated.

† A key early figure here is Philo Judaeus, who identified the Greek idea of the cosmic *Logos* with the Hebrew notion of the "Word" (*dabhar*) of God. In John's Gospel,
this "Word" is identified with Christ as the second person of the Trinity. The synthesis of Judeo-Christian and Greco-Roman elements continues at an accelerated
rate with the early Church Fathers and into the full flowering of Medieval theology
and early modern philosophy.

Planetary Era, that the Western (and predominantly Christian) worldview experiences any sustained encounters with cultures outside of the Abrahamic and Hellenistic matrix from which it emerged. The greatest initial impressions are made through contact with the Aztecs in the far West and the Chinese in the far East. It is the Christian missionaries who are the first to describe and reflect upon these cultures in any depth. It is some of these same missionaries—Bartolome de las Casas in Chiapas and Mateo Ricci in China are the most illustrious examples—who become the first apologists for the intrinsic value of these cultures.* By the turn of the eighteenth century, drawing from the missionary accounts, Leibniz argues for the compatibility of Christianity with the natural theology of the Chinese, while later in the century Voltaire extols Chinese civilization to the detriment of the dominant European Christian variety.

It is not until after the French Revolution that Western intellectuals begin to concern themselves with newly translated Hindu and Buddhist texts, and more generally to construct the notion of "the East." Then as now, their interpretation and significance are colored by the religious and philosophical preoccupations of the times.

Hegel is no exception. In contrast to the generally enthusiastic estimation of German intellectuals of the period (epitomized by the Romantic, Friederich Schlegel), Hegel's portrayal of Hindu philosophy and religion is largely negative (as was the case in most of the sources available to him). On the other hand, as Hodgson notes, his presentation of the Buddhist notion of nirvana—" although couched

---

* And not only the cultures, but the people themselves—witness Bartolome de las Casas' passionate defense of the "Indians" as full humans with immortal souls against the specious arguments of Juan Gines, who argued on Aristotelian grounds that they were not fully human and thus could be treated as slaves. Though Bartolome de las Casas persuaded the theologians who presided over the debate, it was Sepulveda's position that was largely followed in the Indies (see oregonstate. edu/instruct/phl302/philosophers/las_casas.html (retrieved April 4/16/08).

in Western ontological categories—brings Hegel to a defense of Oriental pantheism" against its Christian theological critics.[89]

This is not the place to enter into a discussion of the distortions or other shortcomings of Hegel's analysis of non-Christian religions. It is important to note, however, that with Hegel we have the first comprehensive and systematic attempt at a comparative history of world religions. As is the case with each part of his system (and of the system as a whole), Hegel's philosophy of religion is organized according to the same triphasic pattern that informs the present account of the birth and transformation of the Planetary Era. Hegel considers not only the East, but also what we would call indigenous or Earth-based religions, under the first moment (identity) of "immediate religion" or the religion of nature. The second moment (difference) includes Greek religion and Judaism, both of which effect an "elevation of the spiritual above the natural." The third moment (new identity) corresponds to Christianity as the "consummate" or "absolute" religion, in the sense that its central symbols of Trinity and Incarnation point to the complex synthesis or mediation of the fundamental opposites (nature and spirit, universal and particular, divine and human, and so on).

An open question is the degree to which certain Eastern traditions of which Hegel was unaware, or that he perhaps misunderstood, might also give adequate expression to the notion of the Absolute as the identity of identify and difference, or what, following Aurobindo, we might call the integrally non-dual. Apart from Aurobindo himself—who was heavily influenced by Hegel—I would point especially to the "qualified non-dualism" of Ramanuja and the affiliated tradition of *acintya bhedābheda* ("inconceivable oneness and difference"), and in Buddhism to the Huayan or Kegon schools (from which many New Paradigm writers borrow the analogy of Indra's net, in which each jewel holographically reflects all the others) and the increasingly popular Tibetan traditions of Dzogchen

("Great Perfection"). These Buddhist traditions of non-duality all draw from the classic expression of the "perfection of wisdom" (*prajnaparamita*) in the Heart Sutra: "form is emptiness and the very emptiness is form."

As we shall see in chapter 12, Ken Wilber for one considers some of these traditions to be more realized (more "consumate" or perfected, in Hegel's language) than Christianity. In my estimation, the question is inherently undecidable in anything but personal terms. We have already seen how Eastern versions of the fundamental pattern, though applied to the realm of psycho-spiritual development, have not (or were not in their original contexts) applied to history as such or the evolution of consciousness in general. Though the Christian inflection of the triadic Great Code may not be more "perfected" or realized in any ultimate sense, it does seem to be unique in its fusion with the linear ufolding of time as ("salvation") history. In any case, from my point of view, Christianity is not to be conceived as the absolute Omega, but as the already syncretic, hybridized, or complex Alpha from and through which the West continues to evolve into a post- or meta-Christian Planetary Era, the still-emerging soul of which is the true, if yet uncertain and perhaps never fully attainable, Omega point, the intuited goal of our long-anticipated homecoming.

FROM THE NEW ENLIGHTENMENT
TO THE SIXTIES COUNTERCULTURE

J UST as the organicist and enchanted qualities of the cosmos evident at the birth of the modern period eventually give way to the mechanistic and disenchanted cosmos of the Cartesian–Newtonian paradigm and the worldview of the Enlightenment, so the mini-axial age of flourishing Romanticism and Idealism gives way, toward the mid nineteenth century, to what Baumer has called the "new enlightenment."[90] The major figure in Britain here is John Stuart Mill, whose philosophy of utilitarianism (making pleasure and pain the primary motives of human behavior) and associationism (the contents of consciousness, however complex seeming, are the product of the mechanical interaction of simple, sense-derived data) is the direct descendent of the mechanistic tradition beginning with Hobbes and Locke and running through Hartley and Hume to Mill's mentor Jeremy Bentham. In Germany there are the so-called left-wing Hegelians—including Feuerbach and, most famously, Marx—who "stood Hegel on his head" and, taking up the cudgels of the radical Enlightenment and the French Revolution, articulated renewed and powerful forms of atheistic materialism in the interests of "a ruthless critique of everything existing."[91]

The third main expression of the new enlightenment is positivism, which, though originating in France, was, like Marxism, to have a major influence across the western world throughout the nineteenth and twentieth centuries. The father of positivism and of modern sociology, Auguste Comte (1798–1857), is an intriguing figure whose life and teachings are considerably more complex and textured than one might gather from the standard associations to the word *positivistic*.* The ontologically flat and spiritually impoverished character of these associations can be gathered from the following utterances of Claude Bernard from what could be called the "credo of positivist science":

> If our feeling constantly puts the question *why*, our reason shows us only that the question *how* is within our range...we must always seek to exclude life entirely from our explanations of physiological phenomena.... Life is nothing but a word which means ignorance.... Science should always explain obscurity and complexity by clearer and simpler ideas.[92]

In contrast to his followers, though he also looks to the cultivation of "the positive state" where "the mind has given over the vain search after Absolute notions, the origin and destination of the universe,"[93] Comte nonetheless advocates a new cult of Humanity to elicit our "sympathetic instincts" (in place of utilitarian "calculations of self-interest"). Despite the glorification of Newtonian science and the general rejection of final causes, Comte finds "the great conception of Humanity" to be the center "toward which every aspect of Positivism naturally converges." Even more surprisingly, Comte tells us that, through its appeal to the heart as well as to the intellect, it becomes apparent that it is "the principle of Love upon which the whole system rests."[94]

Comte is most remembered, however, for his "fundamental law" of three stages to the evolution of worldviews.† "The law is this,"

---

* *Positive* here refers to what is empirically and scientifically grounded. The term is derived by contrast with the logical process of establishing conceptual distinctions through "negation."

† As Reeves and Gould (62) point out, Comte was aware of Joachim and his followers' notion of the Three Ages and possibly even drew inspiration from it. "In the thirteenth century," writes Comte, "a more radical attempt, initiated by the pious

states Comte: "that each of our leading conceptions—each branch
of our knowledge—passes successively through three different theoretical conditions: the Theological, or fictitious; the Metaphysical,
or abstract; and the Scientific, or positive."[95] Despite its three stages,
Comte's linear and monochromatic scheme is clearly a truncated
expression of the fundamental pattern we have been exploring. His
Theological stage corresponds to the first phase of the larger arc, but
his second and third stages—the Metaphysical and Scientific—are
both inflections of the *second* moment (differentiation) of the more
fundamental pattern (see figure 1). While Comte's utopian vision of
a world united through the cult of Humanity—which itself is said
to bring about a marriage of head and heart, of Reason and Love—
betrays the deeper incarnational impulse within the positivist aspiration, the categories from the Enlightenment worldview that Comte
has inherited are simply not equal to the vision's promise. There is no
genuine third moment, and thus his scheme does not allow for a full
reading of the Great Code that, as we have seen, permits us to situate
both the positivist movement itself as well as the anticipated new age
of which it saw itself the harbinger.

Comte's mentor, Saint-Simon (1760–1825), actually proposed a
scheme for understanding the history of worldviews that, though
involving only two factors instead of three, captures something of
the dialectical character of the fundamental pattern we have been
working with. According to Saint-Simon, history is marked by the
alternation of "organic" or constructive and "critical" or negative periods. While the philosophy of the eighteenth century had been critical,
that of the nineteenth century would be organic[96]—which it was for
a while, as we have seen, until the Romantic and Idealist movements

---

utopian [Joachim]whom Dante installed in Heaven for his prophetic spirit, was
realized by Bonaventure's worthy predecessor [John of Parma].... His book [*The
Eternal Evangel*], unknown today, was then the mouthpiece of the highest aspirations, and strove nobly to elevate the status of the third person of the Trinity in
order to inaugurate the reign of the heart." (my translation).

were supplanted by the new enlightenment (a new "critical" phase). Despite, however, the obvious resonances between his "organic" and the first and third moments of the fundamental pattern, and between his "critical" and the second moment, there is no indication that Saint-Simon was aware or made use of the more complex and generative potential of the triphasic fundamental pattern.

The most significant figures after Comte (I have already mentioned Mill and Marx) to carry forward the spirit of the new enlightenment are Darwin (1809–1882), Freud (1856–1939), and Skinner (1904–1990). It is the essentially mechanistic and materialistic account of evolution so forcefully articulated by Darwin and his followers that captures the imagination of successive generations of scientists and intellectuals, most of whom, to the extent that they are even aware of its existence, reject out of hand the earlier, spiritually oriented evolutionary worldview of the Idealists and *Naturphilosophen.* Freud, as we know, though clearly influenced by the Romantic worldview—most obviously in the central role accorded to the notion of the unconscious—made no secret that he fully expected the life of the psyche, which he did so much to illuminate, eventually to be accounted for in purely materialistic and mechanistic terms. As for Skinner—whose behaviorism would, along with Freud's psychoanalysis, dominate psychology and psychiatry for most of the twentieth century—mind or psyche, in all seriousness, is simply rejected out of hand as a needless fiction. With the supplanting of both psychoanalysis and behaviorism by psychopharmacology, and with the rise of the powerful biotechnology industry, it would appear that considerable progress has been made toward the realization of the positivist's dream of banishing spirit, mind, and life from the discourse that defines the scientific worldview.

---

*   This is equally true for Herbert Spencer (1820–1903). Though in some ways more organic-holistic than Darwin (who, incidentally, borrowed the phrase "survival of the fittest" from Spencer), his vast evolutionary synthesis remained fundamentally mechanistic (in its "laws") and, under the cover of a metaphysical/religious agnosticism, thoroughly disenchanted.

There have, however, been significant countercultural impulses since the eclipse of the Romantic and Idealist movements. We have seen how Romanticism and Idealism arose at the height of the Enlightenment. Similarly, at the high point of the new enlightenment surrounding the threshold of the twentieth century (ca 1880–1920), which was otherwise flush with the most inflated forms of the Cartesian–Newtonian paradigm, we find a cluster of related figures and movements that manifest varying degrees of resonance with the preceding Romantic/Idealist movements (and the more esoterically inflected expressions of the birth of the modern period). Though perhaps not quite as coherent as the two preceding countercultural periods, the cluster does point to the completion of a new cycle and a third fractal repetition of the larger arc.

For an intuitive sense of the transformations underway during this period, one merely has to consider the revolution that took place in the arts. In painting, immediately following the light-liberating effect of the impressionists, we have the emergence of post-impressionism (Cézanne, Van Gogh), fauvism (Matisse), expressivism (Kandinsky), cubism (Picasso, Braque), and the seeds of surrealism (Breton), all of which challenge the materialistic and literalistic "single vision"[*] of the dominant intellectual culture. In literature, the constraints of nineteenth-century realism give way to the experimentation of the modernists, most notably with the introduction of "stream of consciousness" writing (Proust, Gertrude Stein, Joyce, T. S. Eliot). In music, similarly, this period sees the final flowering of late Romanticism (Wagner, Brahms, Tchaikovsky) and the definitive mutation of the Western sense of rhythm and tonality, initiated by Beethoven during the high point of the previous cycle, with the works of Scriabin and especially those of Stravinsky (*The Firebird*

---

[*]   From Blake: "Now I a fourfold vision see,/And a fourfold vision is given to me; / 'Tis fourfold in my supreme delight / And threefold in soft Beulah's night/And twofold Always. / May God us keep / From Single vision & Newton's Sleep!"

and *The Rite of Spring*) and Arnold Schoenberg (for example, his *Das Buch der Hängenden Gärten*).

The most obvious challenge to the Cartesian–Newtonian paradigm arises toward the middle of this period with the birth of the new physics. Just after Lord Kelvin announced the end of physics (as just about all of the outstanding problems seemed to have been solved), Max Planck discovers the quantum and Einstein lays the foundations for the theory of relativity. I shall have more to say about the new physics in chapters 11 and 12. Here all we need note is that, after three centuries of schooling in the principles of the mechanistic worldview, the very foundations of science begin to offer up a view of the cosmos that is much more compatible with that of the forgotten or abandoned Romantic and Idealist philosophers of nature and the earlier Renaissance esotericists. In contrast to the more rigid dualism characteristic of the dominant mechanistic worldview, the new physics and the cosmology that arises out of it present us with the idea that such fundamentals of physics as matter and energy and space and time (and other more specific pairs, such as wave and particle, position and velocity, locality and non-locality) are related to each other in dynamic, complementary, and even paradoxical ways. As David Bohm will later put it, the new physics suggests that the nature of the cosmos is best described not as a machine, but as "unbroken wholeness in flowing movement."[97] Earlier in the twentieth century, Alfred North Whitehead will propose a comprehensive philosophy of *organism* (also known as *process* philosophy) based on the findings of the new physics.

Thompson focuses on 1889 as the year that signaled the birth not only of the new "Complex Dynamical Mentality" in science, but also that of the new planetary culture. Pointing to the relation of the new science to the soon-to-follow revolution in the arts, on the one hand, and forward to the emergence of complexity theory in the later twentieth century, he writes:

1889 is the magic year for the emergence of the new Complex Dynamical Mentality, for that is the year Poincaré discovers the homoclinic tangle (or the three-body problem in physics) and discovers that the solar system is not the orderly elliptical system of Kepler, but a chaotic system. 1889 is as well a good year to date the emergence of a new planetary culture through the agency of 'the City of Light', because that was the year of he Universal Exhibition in Paris.... So let us date the emergence of planetary culture with this year of 1889. Poincaré's works would influence Einstein, Picasso and Kupka, and they would also initiate new directions in the mathematics that followed with the catastrophe theory of René Thom's work in the 1960s in Paris, the Chaos Dynamics of the 70s and 80s in the United Sates, and the foundation of the Santa Fe Institute for the Sciences of Complexity in the 1990s. What had been only an idea in the mind of a mathematical genius in 1889 was a new global scientific culture by 1989.[98]

We have seen how the organism, or life, rather than the machine, is the preferred root metaphor for the esoteric and Romantic/Idealist traditions. It is during this period surrounding the threshold of the twentieth century that we see a wave of figures for whom a non- or anti-mechanistic view of life is once again the preferred root metaphor. To mention only the most prominent, there is Nietzsche (drawing from the late Romantic, Schopenhauer), for whom all of reality is conceived as a manifestation of Life as "Will to Power" or "self-overcoming." The embryologist Hans Driesch proposes a new scientific theory of *vitalism* with his notion of "entelechy," the inner principle or directive force responsible for the emergence, development, and maintenance of organic forms. Henri Bergson's related notion of *élan vitale* as the driving force of "creative evolution" (the title of his most famous book) gains a wide reception, not only in Europe, but in America as well with William James, whose popular idea of the "stream of consciousness" was directly influenced by his reading of Bergson. It is also during this period that the biologist, artist, and evolutionary philosopher Ernst

Haeckel, who coined the term *ecology,* develops an influential world-view based on the intuition of the wholeness of the living cosmos.

At the same time that the foundations of physics are shaken by Planck and Einstein, the dominant view of consciousness or the nature of human subjectivity is challenged with the emergence of depth, and what will later be called transpersonal, psychology (more on transpersonal psychology in chapter 11). In contrast to the Cartesian and Enlightenment ideal of autonomous reason, Nietzsche, Freud, and Jung reintroduce the Romantic notion of the *unconscious* as the deeper or wider dimension of the psyche, a dimension that is grounded not in reason but in life. Instead of the unconscious, James will call it the *transmarginal field* of consciousness, while his friend and colleague, Frederick Myers, will call it the *subliminal self.* For Jung, James, and Myers, this deeper and wider dimension of the psyche is also a source of spiritual or transpersonal potentials. It is for the scientific investigation of such potentials that the Society for Psychical Research is founded at this time (1882) and for which both Myers and James will serve as presidents.˙ Already by the end of the first phase of the Society's activities following the death of Myers in 1901, an enormous amount of data on what later came to be called "psi" phenomena (telepathy, psychokenesis, apparitions, apparent communication with the dead) will have been accumulated. This data, and the theoretical reflections that they generated, constituted a direct challenge to the mechanistic paradigm of the reigning new enlightenment.

Shortly before the founding of the Society for Psychical Research, Madame Blavatsky and associates founded the Theosophical Society

---

* Although Freud maintained a skeptical interest in telepathy (a termed coined by Meyers) throughout his life, early on he repudiated the esoteric and spiritual-metaphysical dimensions of the Romantic worldview in favor of the materialistic orientation of the new enlightenment. Jung, by contrast, wrote his doctoral dissertation on the productions of a spiritualistic medium and developed his theory of the unconscious in a way that was sympathetic to what Freud derisively characterized as the "black tide of mud" of the occult. See, in this connection, Charet (1993).

(1875). During this period surrounding the threshold of the twentieth century, both of these societies attracted the support, or at least
the sympathetic interest, of a large portion of the world's leading
scientists and intellectuals. Following the initial preoccupation with
mediumistic phenomena, by 1896 Blavatsky stated that the goals of
the Society were: "1) To form a nucleus of the Universal Brotherhood
of Humanity, without distinction of race, creed, sex, caste or colour.
2) To encourage the study of comparative religion, philosophy and
science. 3) To investigate the unexplained laws of Nature and the
powers latent in man."[99] Hanegraaff notes that "'science' meant the
occult sciences and philosophy the *occulta philosophia*,…the laws of
nature were of an occult or psychic nature, and…comparative religion was expected to unveil a 'primordial tradion' ultimately mod-
eled on a Hermeticist *philosophia perennis.*"[100] The importance of the
Hermetic tradition to the theosophical movement (especially at its
origins) links it directly to the earlier Romantic and Idealist movements, as does its opposition to the dominance of the mechanistic
paradigm in science.

   With respect to the esoteric, occult, or perennial philosophical
spirit of the times, two representative events are worthy of note. The
first, though purely symbolic, is the erection, on the day of the Holy
Spirit, 1889, of a statue of Giordano Bruno in the Roman piazza where
he was burned at the sake in 1600 (at the beginning of the second cycle).
The project was sponsored by the Freemasons and the international
planning committee included Victor Hugo, Ernest Renan, Ernst
Haeckel, Hubert Spencer, and Hernrik Ibsen. Thousands participated
in the ceremonial unveiling. Pope Leo XIII was scandalized.

   The second event is also symbolic of the period in question,
though it also had a significant impact on the countercultural
stream. The event was the World Parliament of Religions, held
in Chicago in 1893. Conceived as the first official forum for inter-
religious dialogue, the Parliament featured the encounter between

East and West (though indigenous traditions were not included, some new religious movements were, including Spiritualism and Christian Science). The highlight of the event was the speech by Swami Vivekananda, disciple of the Hindu saint, Ramakrishna. The speech introduced a wide audience to the kind of philosophical Hinduism (Vedanta) that would play such a central role in the later counterculture(s). During this extended stay, Vivekananda made a deep impression on William James, Josiah Royce, and other members of the Harvard faculty, founded the Vedanta Society of New York (1896), and in a later visit (1898) founded a monastic community near San Francisco.*

To summarize, the new cycle—the third fractal repetition of the larger arc—can be represented in the following diagram.

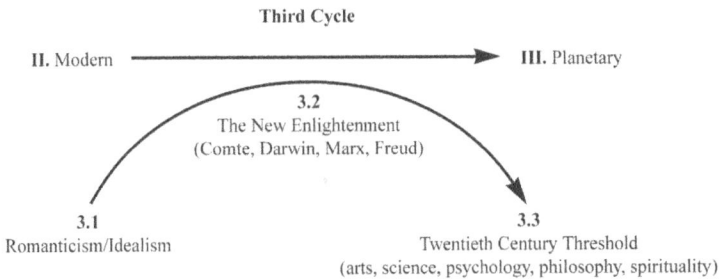

Third Cycle

II. Modern ⟶ III. Planetary

3.2
The New Enlightenment
(Comte, Darwin, Marx, Freud)

3.1
Romanticism/Idealism

3.3
Twentieth Century Threshold
(arts, science, psychology, philosophy, spirituality)

---

* Thompson focuses on 1889 as the year that signaled the birth not only of the new "Complex Dynamical Mentality," but also that of the new planetary culture. "1889 is the magic year for the emergence of the new Complex Dynamical Mentality, for that is the year Poincaré discovers the homoclinic tangle (or the three body problem in physics) and discovers that the solar system is not the orderly elliptical system of Kepler, but a chaotic system. 1889 is as well a good year to date the emergence of a new planetary culture through the agency of 'the City of Light', because that was the year of he Universal Exhibition in Paris.... So let us date the emergence of planetary culture with this year of 1889. Poincaré's works would influence Einstein, Picasso and Kupka, and they would also initiate new directions in the mathematics that followed with the catastrophe theory of René Thom's work in the 1960s in Paris, the Chaos Dynamics of the 70s and 80s in the United Sates, and the foundation of the Santa Fe Institute for the Sciences of Complexity in the 1990s. What had been only an idea in the mind of a mathematical genius in 1889 was a new global scientific culture by 1989" (Thompson, 2004, 45–47).

Despite the radical transformations in the arts, science, psychology, philosophy, and spirituality that took place around the threshold of the twentieth century, however, the dominant worldview remained that of the positivistic and mechanistic (new) enlightenment. Still, the presiding mood of the zeitgeist on either side of the threshold is markedly different. In place of the unclouded optimism of the previous century, the West, beginning with World War I and soon to be followed by the Great Depression, the rise of fascism, and the catastrophe of World War II, now enters what Baumer has called the "age of anxiety."[101] If we take the twentieth century threshold as the first moment of a new, third cycle, the age of anxiety that follows would correspond to its second moment.

The idea of the death of God, which Nietzsche had announced toward the end of the previous century, and which now seemed confirmed in the course of world events, perhaps best epitomizes the depth of this anxiety. As Baumer notes, however, this period also saw "the death of man and the death of Europe; in fact, the death—or at least the toppling—of all the great modern idols: not only God and man, but also reason, science, progress, and history."[102] The great twentieth-century theologian Paul Tillich summed up the prevailing mood toward the end of this period with the observation that the West was gripped with the "threat of spiritual nonbeing" in the form of an overarching "anxiety of emptiness and meaninglessness."[103]

The end of the Second World War coincided with the birth of the nuclear age. Shortly thereafter, the Cold War fragmented the world, already (and still) divided into north and south, into left and right. Both feeding and feeding upon the growing anxiety of emptiness and meaninglessness, the industrialized north and west plunged ever more frenetically into the mass delirium of industrial growth society that, as Theodore Roszak characterized the situation in 1969, "after ruthlessly eroding the traditionally transcendent ends of life, has concomitantly given us a proficiency of technical means that now oscillates absurdly between the production of frivolous abundance and the production of genocidal munitions."[104] Roszak wrote these lines in the opening chapter to his classic book *The Making of a Counter Culture*, where, drawing from the reflections of French social philosopher Jacques Ellul, he outlines the pervasive reach of what he terms *the technocracy*. The technocracy can be defined as "that society in which those who govern justify themselves by appeal to technological experts who, in turn, justify themselves by appeal to scientific forms of knowledge" and to "the dictates of industrial efficiency, rationality, and necessity."[105] By "science" and "rationality" one should read the scient*ism* and primarily *instrumental* reason (Hegel's *Verstand* or "understanding") of the new enlightenment, whose cadre of experts by this time had become the handsomely funded lackeys of the (techno) military industrial complex.

It is perhaps, as Roszak suggests, as protest against the technocracy and the correlative psychosocial and spiritual wasteland* of the postwar period that we can best comprehend the rise of the sixties counterculture. "The revolution which is beginning," we read from a manifesto pinned to the entrance of the Sorbonne in May, 1968, "will call in question not only capitalist society but industrial society.... The society of alienation must disappear from history. We are inventing a new and original world. Imagination is seizing power."[106] One can discern two principal inflections of the sixties counterculture that, though sharing an opposition to the status quo, are themselves both complementary and in a certain sense antagonistic to each other. The two inflections can be distinguished by a preponderance of what Jung defined as extraversion or introversion and, with respect to the San Francisco Bay area, which in many ways functioned as the spiritual epicenter for the movement as a whole, might be associated more with either the University of California in Berkeley or the Haight-Ashbury district in San Francisco, respectively. On one

---

* The title of Roszak's next major work is *Where the Wasteland Ends* (1972).

side, there is a call for freedom and justice through organized protest against dominant social and political power structures and their
agents or enforcers: university administrations, government offi-
cials, the police, the military, banks, corporations, the church. The
paradigmatic events in this complex movement are the student protests, from Berkeley to Kent State in the United States, but including non-American cities as well—most notably Paris and Mexico
City. Inspired by and taking up the Southern civil rights struggle
and adopting its techniques, this inflection of the counterculture
opposed the Vietnam War and championed the cause of women's
rights, free speech, the environment, and gay liberation.

The other inflection, though generally sharing the same causes—
and keeping in mind that many individuals were capable of embracing both simultaneously—preferred to focus its energies on alternative modes of being and the "expansion of consciousness," on
cultivating those values and qualities that the dominant worldview
either neglected, diminished, or actively suppressed: peace and love,
communitarian and even communal lifestyles, ecological awareness,
an experimental esthetic, nonordinary states of consciousness, liberation of sexuality, the spiritual quest, the reenchantment of the
world. "We grasp the underlying unity of the counter cultural variety," writes Roszak,

> if we see beat-hip bohemianism as an effort to work out the per-
> sonality structure and total life style that follow from New Left
> social criticism. At their best, these young bohemians are the
> would-be utopian pioneers of the world that lies beyond intellec-
> tual rejection of the Great Society. They seek to invent a cultural
> base for New Left politics, to discover new types of community,
> new family patterns, new sexual mores, new kinds of livelihood,
> new esthetic forms, new personal identities on the far side of
> power politics, the bourgeois home, and the consumer society.
> When the New Left calls for peace and gives us a heavy analysis
> of what's what in Vietnam, the hippy quickly translates the word

into shantih, the peace that passes all understanding, and fills in the psychic dimension of the ideal.[107]

Also, writing in 1969 during his extended sojourn in California, Edgar Morin notes the complexity of this "exuberant movement, which is at once formless and multiform, embraces both psychedelia and politics, sexual topics and mystical ones, seeks to establish its identity in a revolution that is sometimes social and sometimes individual (or both at the same time), and encompasses a range of people from the hippie to the militant campus Leftist."[108]

This new cycle—the fourth fractal repetition of the larger arc—can be summarized as follows.

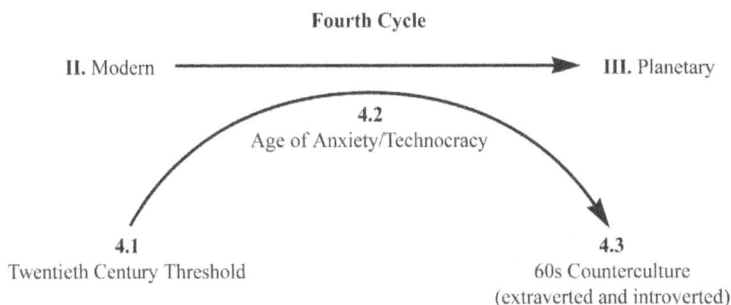

Fourth Cycle

II. Modern ————————————————▶ III. Planetary

4.2
Age of Anxiety/Technocracy

4.1
Twentieth Century Threshold

4.3
60s Counterculture
(extraverted and introverted)

There is a profound resonance between the more introverted inflection of the counterculture and the Romantic spirit of the early nineteenth century. Anticipating the liberational, experimental, and experiential idealism of the sixties, there is, for instance, the fascinating relationship between August Schlegel, Schelling, and Caroline Michaelis Böhmer Schlegel Schelling, whose string of last names tells much of the story;[109] the utopian farm that the young Coleridge and Robert Southey planned to start in America; the influence of opium on the worldview of Coleridge and Thomas de Quincey; the openness on the part of most of the Romantics to the supernatural, to what would come increasingly to be linked to the notion of the unconscious, and to the links between the mind or consciousness and the deeper realms of meaning traditionally associated with mystical religion and occultism.[110]

By contrast, the other, more extroverted, inflection manifests more obvious ties to the spirit of revolutionary France before the Terror (and to the radical sectarians of the earlier British revolutionary period), though individuals with this dominant attitude would most likely share many of the same concerns as their introverted counterparts and even, under the right circumstances, adopt their more introverted attitude. The shared elements between this inflection of the counterculture and the spirit of the French Revolution and of the earlier British revolutionaries include a passionate rejection of received authority: the King, on the one hand, the President on the other; the heightened and articulate consciousness of inalienable rights: particularly, the right to free speech and the right to organize; a struggle to liberate the oppressed: from the urban and rural poor to debtors and slaves. (Slavery in France was abolished in 1791, in 1794 in the French colonies; the constitution of 1794 also abolished imprisonment for debt—colonial slavery, however, was reintroduced by Napoleon in 1801.)

Looking back at the sixties counterculture as a whole, one can see how its position as the third moment in this new cycle—the fourth fractal repetition of the larger arc—means that it is defined not only, or principally, as counter to the worldview that precedes it (the twentieth century continuation of the new enlightenment as second moment). In its more mature or integral incarnations, at least, it also seeks to integrate the previous moments' best or noblest elements into a more comprehensive, organic, or holistic alternative. From the earlier Enlightenment and Revolutionary period, it appeals to the ideals of freedom and justice. From the Romantic/Idealist period, especially in its more introverted inflection, it takes up the esoteric/occult concern with consciousness (and the *un*conscious), meaning, spirit, and an enchanted cosmos. This more integral character of the third moment (recall Hegel's notion of "sublation") is equally apparent in the case of the "theosophical enlightenment" (to use Joscelyn Godwin's phrase)[111] of the preceding cycle at the threshold of the twentieth century and in the case of German Idealism in the cycle before that. Both of these prior third moments sought to preserve the Enlightenment ideal of Reason while incorporating the depth, intuitive, aesthetic, and spiritual insights of the esoteric tradition that inspired poets and philosophers alike (see Hanegraaff).* Similarly, as we saw in chapter 5, the founders of both modern science and of the Reformation did not simply reject, but built upon, the critical, mental culture of medieval scholasticism as they reached back to even earlier sources of inspiration (the Hermetic tradition and primitive Christianity, respectively).

Concerning the sixties counterculture, Charlene Spretnak astutely notes that it is ironic that it "was dismissed as *romantic* even though its ignorance of the Romantics was almost total."

> Except for frequently heard references to William Blake, the fact that a much earlier wave of malcontents had mounted a similar rejection of the ideologies of modernity *and* had arrived at a number of extremely relevant alternatives went unnoticed.... In rejecting the reductionism of rationalism, the counterculture was so deeply anti-intellectual that it forfeited access to its own history.[112]

Perhaps rather than, or at least along with, some such anti-intellectualism, a significant factor in the surprising ignorance among counterculturalists of their Romantic predecessors is the degree to which the contributions of the Romantics (and Idealists, not to mention the earlier radical figures and movements we have considered) were successfully expunged from (or minimized or distorted in) the dominant accounts of the "progress" of Western philosophy and

---

* "The cultural primacy of the esthetical in the Italian Neo-Platonism of the Renaissance," as Gusdorf puts it, "is reborn in post-Galilean romanticism" (Gusdorf, I, 408, my translation).

science. More broadly, the exponential growth in information over the last century, the flooding of the collective consciousness with commercially driven cultural artifacts, and the steady denigration of the past in favor of the new (and preferably disposable) have all contributed to a generalized cultural amnesia. Most of us have forgotten the path that has led us to where we now stand, or even that there was any path at all.

8

A TIGHTENING SPIRAL—A WIDENING GYRE

T HOSE who not only lived through, but actively participated in the
creative unfolding of the sixties counterculture (which actually
extended into the early 1970s) tend to look back on the period with
understandable nostalgia. For many, Wordsworth's lines, recalling
his time in early revolutionary Paris before the Terror, might ring
true for them as well—making allowance, of course, for the filtering
effect of the passage of years and the transmuted form of "emotion
recollected in tranquility"*: "Bliss it was in that dawn to be alive, /
But to be young was very heaven." The heady optimism and unbri-
dled idealism of hippies and new left radicals alike could obviously
not be sustained indefinitely. Even so, the triphasic dialectical pattern we have been tracing has already completed a fifth cycle, and
appears to be ending / beginning another (since the third phase of a
cycle also constitutes the first of the next).

Taken in sequence, the five completed fractal repetitions of the
larger arc trace a tightening spiral, since each cycle is shorter than the
one preceding it: the first is roughly 1,500 years (late first to early seventeenth centuries); the second, around two hundred years (seventeenth
to the nineteenth centuries); the third, around a hundred years (late

---

* This is Wordsworth's formula for the essence of his poetic technique. "Poetry,"
he writes, "is the spontaneous overflow of powerful feelings: it takes its origin from
emotion recollected in tranquility: the emotion is contemplated till, by a species of
reaction, the tranquility gradually disappears, and an emotion, kindred to what was
the subject of contemplation, is gradually produced, and does itself actually exist in
the mind" (*Preface to the Second Edition of Lyrical Ballads*, 1800).

eighteenth to the late nineteenth centuries); the fourth, around eighty years (1889–1968); the fifth, as we shall see, around twenty-five years (1968–1993); and an anticipated sixth, around twenty years (1993–2012). Note that after the second cycle (ending at the turn of the nineteenth century), the overall rate of cycling, though still involving a tightening spiral in terms of length of years, begins to decelerate: the rate of the second cycle is about fifteen percent of the first; the third about forty-five percent of the second; the fourth is about eighty-five percent of the third; the fifth, however, is about twenty-five percent of the fourth (a temporary reacceleration); and the sixth, assuming around 2012, as the ending of the cycle, is about *equal* to the fifth. If, following the first two cycles, the trend had stabilized with each new cycle lasting roughly one-third as long as the preceding, the spiral would have converged into some kind of singularity around 1945. It is true that this year does mark a singular threshold with the end of the Second World War and the beginning of the nuclear age. In terms of the evolution of consciousness with which we are concerned, however—with our focus on the role, within the emerging Planetary Era, of the periodic resurgence of elements from a special set of counter-cultural worldviews—the spiral has continued to cycle. Will the spiral even out into a regular coil, perhaps around a steadily oscillating twenty or twenty-five-year cycle? Will it loosen into increasingly longer cycles, or reaccelerate to converge on some kind of singularity?

We will only know after the fact. Still, given the extremity of the current world situation, and to the extent that the until now counter-cultural spirit of the third/first moments of the preceding cycles seem to point the way to a full actualization of the Planetary Era, the best scenario would be that of a possible singularity. By singularity here I do not mean a full stop, but rather that the spiral would come to approximate a straight line, perhaps with continuing undulations, but no longer any full turns. In concrete terms, this means that planetary culture would have attained a state of sustainably dynamic equilibrium, and would therefore no longer need to be countered in the same fashion.

For now, we will take a closer look at the most recent cycle, the fifth fractal repetition of the larger arc. If we take the sixties counterculture as the new first moment, the new second can be said to have begun (in the United States at least, which, as the dominant superpower, has become the central field of action for the trajectory we have been tracing) by the mid-seventies and lasted throughout the presidential administrations of Reagan, Carter, and Bush senior. Some of the elements or qualities this second moment shares with the second moments of the two preceding cycles include a faith in the ideal of progress (here in the form of limitless economic growth); the rhetoric of freedom; consolidation of wealth and secular/global power (in this case in relation to the Soviet Union and the "Third World"); and an essentially mechanistic and instrumental—and therefore disenchanted—relation to the cosmos (nature as mere resource). Unlike the previous two cycles, however, these qualities are now associated with conservative political and religious agendas. By contrast, the Enlightenment worldviews of the eighteenth and nineteenth centuries were coupled more with liberal (eighteenth century) to leftist (nineteenth century) and, as we have seen, even atheist, agendas (in both centuries).

The spirit of this period has been explored by professor of history and religious studies Philip Jenkins in his *Decade of Nightmares: The End of the Sixties and the Making of Eighties America.*[113] In the decade in question—which Jenkins bookends with the Fall of Saigon in 1975 on one end, and Reagan's second term in 1986, on the other—we see a pervasive shift in the collective psyche from the optimistic and communitarian idealism of the sixties counterculture to a reactionary and fear-based culture obsessed with the threat of various kinds of religious and moral evil. This period sees Reagan's Manichaean characterization of the Soviet Union as the "evil empire," the beginning of the "war on drugs," OPEC's "stranglehold" on American oil, and "America held hostage" in the Iran hostage crisis. Jenkins also points to the preoccupation in the mass media with serial killers (the phrase was coined in the mid-1970s), ritual child sex abuse, and the new threat of international terrorism. Though focusing on the American scene, Jenkins points out that the rise of the new religious and political Right in this period—the Moral Majority was founded in 1978—was part of a worldwide surge of religious fundamentalism: the creation of the Likud party in Israel, the rise of the Janata (precursor of the BJP) party in India, the Iranian Revolution, the spread of sharia law across Asia, Indonesia, North Africa, and Nigeria.

Three iconic world-historical developments signal the shift to the third moment of this fifth fractal cycle: the spread of the Internet and the invention of the World Wide Web in 1989; the fall of the Berlin Wall in the same year; and the fall of the Soviet Union, and thus the end of the Cold War, in 1991. Together with such events as the forming of the European Union or E.U. (1993), the establishment of the World Trade Organization or WTO (1995), and the creation of the Intergovernmental Panel on Climate Change or IPCC (1988), these developments point to a critical new phase in the unfolding of the Planetary Era, a phase in which the idea and fact of the planetary (or global, as it is more often called in this period) begins to receive unprecedented and sustained attention.

In the sciences, this period sees the rise to dominance in theoretical physics of superstring theory, a new popular awareness of Big Bang cosmology (epitomized by Hawking's 1988 bestselling *A Brief History of Time*), the discovery of dark matter and energy, and growing nonspecialist interest in chaos and complexity theory. As significant as these theoretical developments might be, however, the period is remarkable for a series of technological and experimental breakthroughs, including the stunning revelations of the deep cosmos by the Hubble telescope, the beginning of the Human Genome Project, the first animal cloning (Dolly the sheep), the discovery of the first extra-solar planets, the construction of the first International Space Station, and the full deployment of the Global Positioning System

(GPS). All of these suggest a fundamental reorganization of the human relative to the whole or totality—life, the whole physical cosmos, and perhaps most concretely, the planet.

On the American scene during this period, the shift to the third moment of this cycle is most apparent in the early years of the Clinton presidency and the coming into their own of the baby boomers, the elder of which had grown up during the period of the sixties counterculture. Despite the prominent concern with material comfort and a continuation (though arguably tempered) of the previous administration's imperialist agenda, these years brought not only the great economic boom associated with the new dotcom industries, but also a resurgence of many of the spiritual values and themes that were prominent in both the sixties and in the preceding proto-counterculture of the Romantic/Idealist era, and indeed during the Renaissance at the origins of the modern.

Anticipating and helping catalyze the New Paradigm thinking that would flourish in this period (considered in chapter 11), William Irwin Thompson wrote in *Pacific Shift* (1986): "It would appear that we are more in a period like the Renaissance than like the eighteenth century: a period of new intuitions in poetry, art, and philosophy...."[114] A few years later, in *Gaia 2: Emergence: The New Science of Becoming*, Thompson made the (sadly) overly optimistic prediction that "the nineties is the period in which we shall have to shift from economics to ecology as the governing science of the modern world."[115] In his retrospective of this period, Richard Tarnas points to

> the pervasiveness and intensity of contemporary Western interest
> in Buddhism, Hinduism, and Taoism, in meditation and mysticism,
> in esoteric traditions and mythology, in Jungian and archetypal
> psychology, in transpersonal theory and consciousness research,
> in shamanism and indigenous traditions, in nature mysticism, in
> the convergence of science and spirituality, and in the emergence
> of holistic and participatory paradigms in virtually every field.[116]

Later on in his discussion, Tarnas characterizes the deeper archetypal impulse at work as that of the striving for the conjunction of opposites, a central concept, as we have seen, behind not only Jung's psychology but also the worldview of the Romantic/Idealist era. Echoing the guiding preoccupations of the third moments of the previous cycles already considered, Tarnas describes this particular resurgence in terms of a

> collective awakening of a spiritual and existential desire to merge with a greater unity—to reconnect with the Earth and all forms of life on it, with the global community, with the cosmos, with the spiritual ground of life, with the community of being. This archetypal impulse has been visible as well in the new awareness of the *anima mundi*, the soul of the world, the archetypal dimension of life, with the widespread call for a reenchantment of the world—the reenchantment of nature, of science, of art, of everyday life...in the nearly ubiquitous urge to overcome old separations and dualisms—between the human being and nature, spirit and matter, mind and body, subject and object, intellect and soul, masculine and feminine, psyche and cosmos—to discover a deeper integral reality and unitive consciousness. All these tendencies can be understood in terms of the archetypal impulse associated with the *coniunctio oppositorum*, the conjunction of opposites, and the *hieros gamos,* the sacred marriage.[117]

We shall consider various and continuing manifestations of this archetypal impulse in greater detail in part 3.

If the resurgence of countercultural themes in late eighties and early nineties signaled both the end of the previous cycle and the first moment of the next (sixth) one, the new second moment can be said to have arrived in earnest, in the United States at least, with the turn of the millennium and the contested presidential elections of 2000, exasperating an already polarized populace. The elections were soon followed by the traumatizing events of 9/11 and the tragedy of the Iraq war. As with the second moment of the immediately preceding cycle (ca 1975–1986), we saw the re-empowerment of the neoconservative agenda—this time both catalyzed and camouflaged by the so-called global war on terror (which, in the words of activist Vandana Shiva, could more truthfully be described as a "war on Terra"). Under the Bush administration, the United States saw an unparalleled push to dismantle or subvert a wide array of essential social, economic, environmental, and liberal-democratic structures (many of which were strengthened during the previous two countercultural surges), all in the name of "freedom," "liberty," and "Christian values." Under the cover of such doublespeak names as "The Clear Skies Initiative," "No Child Left Behind," "The Patriot Act," and "Operation Iraqi Freedom," this administration sought to secure the U.S. position of global hegemony (the illegal invasion of Iraq), to erase prior gains toward environmental protection (not only through direct legislation, but through illegitimate control of the Environmental Protection Agency), to deregulate the economy (especially in the energy and trade sectors), to justify its use of torture (epitomized in the Abu Ghraib scandal), and ignore fundamental civil liberties (the Patriot Act's suspension of *habeas corpus;* the domestic spying program).

Though the course of the Planetary Era traces the figure of a tightening spiral, it can, as the visionary poet Yeats was able to discern almost a century ago now, also increasingly be described as a "widening gyre." Yeats's poem, "The Second Coming," speaks powerfully to where we now find ourselves. The first stanza deserves to be quoted in full:

> Turning and turning in the widening gyre
> The falcon cannot hear the falconer;
> Things fall apart; the centre cannot hold;
> Mere anarchy is loosed upon the world,
> The blood-dimmed tide is loosed, and everywhere
> The ceremony of innocence is drowned;
> The best lack all conviction, while the worst
> Are full of passionate intensity.

The poem clearly reflects Yeats's perception of the collective European psyche in the time between the two World Wars when it was composed (1920). With the events of 9/11, however, another "ceremony of innocence" was "drowned," or in this case leveled. Many to the left of the preceding U.S. administration (and some even to the left of the current one)—those whose worldview resonates more with the countercultural values of the nineties, the sixties, and the Romantic Era—believe that its leaders did not so much lack conviction as that their convictions were full of a misguided (if not malevolent) intensity. In any case, this intensity is blind to the complex character of the desperate crisis, or series of crises, into which the planet continues to plunge, including devastating climate change and species extinction, a mounting food and energy crisis, and worsening economic and international tensions. There is every indication that, without a drastic change of course, anarchy might indeed be loosed upon the world.*

And yet, at the time of writing, there are signs of a new third moment and the beginnings of a seventh turning of the spiral. We shall consider the spirit of this "Great Turning," as Joanna Macy calls it, in part 3.

The last two cycles, the fifth and sixth, are summarized in the diagrams on the facing page.

* See Robert Kaplan's *The Coming Anarchy* (2001). The subtitle to Kaplan's original article in the *Atlantic Monthly* summarizes well the main argument: "How scarcity, crime, overpopulation, tribalism, and disease are rapidly destroying the social fabric of our planet." See also the grim assessments of James Lovelock (2007).

Fifth Cycle

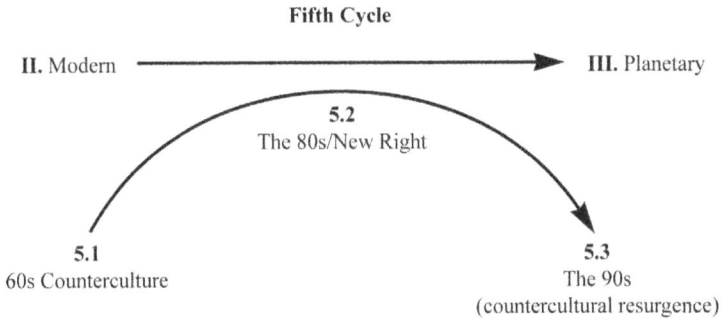

II. Modern

III. Planetary

5.2
The 80s/New Right

5.1
60s Counterculture

5.3
The 90s
(countercultural resurgence)

Sixth Cycle

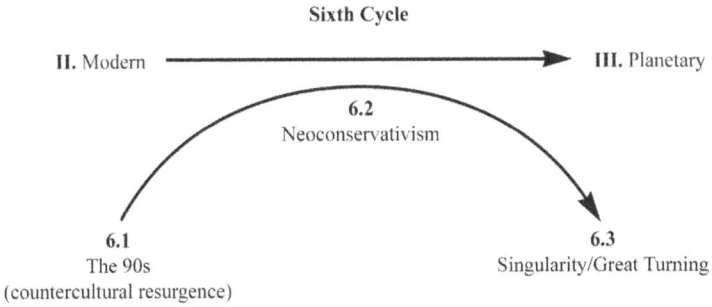

II. Modern

III. Planetary

6.2
Neoconservativism

6.1
The 90s
(countercultural resurgence)

6.3
Singularity/Great Turning

Part Three

Coming Home

"*The onset of the new Spirit is the product of a widespread upheaval in various forms of culture, the prize at the end of a complicated, torturous path and of just as variegated and strenuous effort.*"

—G. W. F. Hegel

"*No one knows what will happen next, such portents fill the days and nights. Years prophetical!*"

—Walt Whitman

"*...this is the largest social movement in all of history. No one knows its scope, and how it functions is more mysterious than what meets the eye.*"

—Paul Hawken

9

LENGTHENING SHADOWS

IN the preceding pages we have seen that, from the beginnings of the Christian era, and increasingly throughout the modern or Planetary Era, the Western mind has been caught up in an extended birthing of a new identity. It has been straddling, or crossing and recrossing, the threshold between the second and third phases of the hero's journey or monomyth. But the runaway character of the modern project has brought us to an unparalleled moment—what the ancients referred to as the *kairos*—and perhaps closer to the end (*eschaton*) that was originally evoked in the biblical version of the monomyth and has been variously re-envisioned by such figures as Joachim, Blake, Hegel, Comte, Nietzsche, Yeats, and Jung. There are at least three definite stances or postures toward our current moment. The first two, which are treated in this chapter, both involve a form of fundamental denial, not necessarily of the urgency of our moment or of the imminence of some kind of end, but of what, from the perspective of the third stance, must be integrated and embraced in order for the moment to pass and for the end to be (at least more) fully actualized. By "end," it should be stressed, I do not mean a literal, apocalyptic terminus as is portrayed in the last book of the Bible, though the current threat to the biosphere and to the viability of human civilization do suggest something very much like this terminus. Rather, I mean "end" in the sense of the deeper *telos*—goal or aim—of the unfolding Planetary Era, indications of which will be considered in the last three chapters.

The first posture could be characterized as involving a form of *regression*, in the sense of a movement that seeks to reinstate a form of identity that was once dominant but that is no longer adequate to the complexity of the current developmental phase or configuration. In the present context, the most obvious and pernicious manifestation of such regression is *militant religious fundamentalism*. What all religious fundamentalisms have in common is a rejection of the forces of secularization that, as we have seen, came to dominance during and following the Enlightenment. This rejection involves a refusal of key elements of modernity, a refusal to pass from the first to the second phase of the larger evolutionary arc—or rather, in the face of a world that has largely already made such a passage, it is an attempt to preserve or reinstate the premodern identity of the first phase. Typical of this prior identity is a naive, pre-critical epistemological orientation. In the case of religious fundamentalism, this orientation tends to be associated with a literalistic reading of sacred texts, which comes into direct conflict not only with critical thinking as such (in the consideration of discrepancies between the three so-called synoptic gospels, for example) but more pointedly with root assumptions of the modern scientific worldview (most notably, with the rejection of the theory of evolution in favor of creationism).

Religious fundamentalism also tends to be coupled with a pronounced authoritarianism. While the authority is nominally grounded in a sacred text, or rather its literalistic reading, real authority rests with the leaders of the group, whose pronouncements do not stop at the question of the authority of scripture, but radiate out to all aspects of life, from the personal (for example, the issues of abortion and gay marriage) to the overtly political (which presidential candidate to vote for). Because authoritarianism inhibits the expression of dissent and of individual differences generally, fundamentalism is also naturally associated with the pressure to conform, on the one hand, and an intolerance of "outside" views, on the other.

All of these traits or tendencies are manifest to the extreme in the case of *militant* religious fundamentalism, where, as we see increasingly in our own times, individuals and groups are led to the clearly demented view that an all-knowing and loving God would rejoice in the slaughter of innocent people. To the terrorist, it is true, the murdered or injured are not "innocent," since their personal identities are nullified by virtue of their participation in the infected collective that is being targeted. Even when truly innocent people are inadvertently targeted, the error tends to be excused on the basis of the greater good that is being realized through the terrorist gesture itself. This line of reasoning is of course the same as that behind the military "doctrine" of "collateral damage." In all three cases, the reality of the individual, of the person, is sacrificed to the narcissistic will of the collective (itself typically embodied in or subordinated to the equally narcissistic will of the leader).

Along with the security that the fundamentalist mindset can give to individuals challenged by the ambiguities or complexities of contemporary life, religious fundamentalism offers a seeming antidote to the disenchantment of the modern worldview. Beneath or beside the authoritarian structure, there is doubtless for many an authentic longing for a world suffused with meaning and purpose, for the recognition of the reality of spirit or transcendence that had characterized the lives of human beings for tens of thousands of years until the recent victory of the metaphysically flattened and seemingly godless worldview of the modern Enlightenment. For many, it is better to live in a split world—simultaneously modern and premodern—than it is to live solely in the spiritual wasteland of late modernity, however "rational" or coherent the latter may see itself as being.

The second stance toward our critical moment in the history of the Planetary Era, and to the end that this moment seems to bring us ever closer to, could be described as involving a kind of *fixation*. Here, instead of attempting to reinstate the form of identity that was dominant during the first phase of the evolution of the Western mind, it is a matter of seeking to maintain or hold on to the modern mindset that largely replaced it and, as we have seen, became the instrument both for the rise of the West and the emergence of the Planetary Era. In many ways this mindset is harder to recognize than the fundamentalist since, as Charlene Spretnak phrases it, "modernity is to us as water to a fish."[118] It is, in other words, a synonym for the still dominant paradigm that, unless both intentionally and studiously resisted, constitutes the parameters and governing categories of our thought, discourse, and general behavior.

We have seen how central both science and the process of secularization are to the modern project. On the science side, the store of knowledge that has accumulated over the past few centuries, from physics, cosmology, and chemistry to biology, ecology, and the social sciences, is staggering both in sheer quantity and in qualitative complexity. Many associated technological breakthroughs, especially in medicine and communications, have enriched the general quality of life for large portions of the population. The process of secularization, for its part, has led to a near global consensus around the guiding social and political ideals of freedom and justice, however variously interpreted and imperfectly implemented. At the same time, and despite their deeper potentials or their organic role in the larger historical process we have been considering, the practice of science and the values associated with secularization have been taken up or co-opted by the corporate-driven techno-industrial complex that has come to dominate the planet during the late modern period.

One sees the mark of this complex in Spretnak's summary of the values of modernity, which includes the following interrelated ideas: the restriction of human being to the sphere of economics (the notion of *homo economicus*) and the related ideal of unlimited economic expansion; the overvaluation of objectivism, rationalism, and reductionism coupled with the rule of mechanistic science; the standardization and bureaucratization of work; the compartmental-ization of education, and indeed of all aspects of life (work, politics, social life, love life, spirituality); and finally, the "preference for com-petition and a dominance-or-submission dichotomy as the structure of all relationships in the spheres of modern life."[119] To summarize, one could characterize these dominant values of late modernity as expressions of a materialistic, power-seeking mentality that, in the pursuit of its ends, imposes a disjunctive or reductionistic logic—the mechanistic paradigm—on all aspects of life.

While I have distinguished between two stances—regression and fixation—toward the challenges of our current moment, the two forces I have highlighted as the leading representatives of these two stances have, in the United States at least, forged an unholy alliance in the service of a common right-wing agenda.* Though religious fundamen-talism and the corporate machine are, from one perspective, rooted in antagonistic phases of the larger evolutionary arc—the religious (first) and the secular (second), respectively—they are both also pathological expressions of the same modern movement of differentiation, which, in these two cases, has led to a pronounced *dissociation*. This dissociation is most apparent in a series of dominance-ordered dichotomies that pits the human against the natural (unlimited economic growth); the few (the so-called developed nations) against the many (the so-called developing nations, the world's poor); the faithful against the "infi-dels." Whether of the fundamentalist, the purely secular-corporate, or the especially virulent hybrid mindset, these powerful and self-serving players in the drama of the current *kairos* are equally blind to what Jung calls the shadow, to the mutually reinforcing feelings of anxiety, fear, hatred, and lust for power that fuel their toxic worldviews. The shadows, moreover, are lengthening, which means that, though they are more powerful, they are also easier to see.

---

* Since I first wrote these words, this agenda suffered a significant setback with the election of the a Democratic administration.

10

GLOBAL SOLIDARITY

IN contrast with the kind of fixation or regression considered in the previous chapter, a potentially transformative engagement with the crises of the Planetary Era is increasingly apparent on two related fronts. These fronts are in direct continuity with the countercultural impetus of the sixties and seventies, and it is likely from here that we can expect the emergence of the next third moment of the spiral journey we have been tracing. The first front, considered in this chapter, is heir to the more extraverted and explicitly activist wing of the counterculture. A major theme of this front is opposition to the corporate-driven push for a so-called free market on a global scale, an opposition that has given rise to the term *anti-globalization*. It is not so much globalization as such that is opposed, however, as it is the policies and practices that seek to subjugate the larger portion of the global community (the "Third" or "Developing" World) and the planet's remaining non-renewable resources.

Instead of "anti-globalization," therefore, many are using George Monbiot's positive phrase "global justice"[120] to describe the movement in question. While one of the first international anti-globalization protests took place in dozens of cities around the world on June 18, 1999, most members of the movement look upon Seattle, November 30, 1999 (known as "N30"), as the founding action. Massive protests successfully disrupted meetings of the World Trade Organization, forcing cancellation of the opening ceremonies. There were more than six hundred arrests and dozens of injuries, despite the generally peaceful behavior of the protestors. There were many more injuries (due largely to police brutality) in Genoa, July 18 to 22, 2001, in the protest against the Group of Eight Summit.* Of the February 15, 2003, events, Hawken notes that it "was the largest coordinated public demonstration in history, with estimates of two million demonstrators in Rome alone."[121]

Despite significant overlap with many of the ideals and objectives of the earlier civil rights, free speech, and New Left movements (with the war in Iraq taking on much the same symbolic role as the Vietnam

---

* Along with Seattle and Genoa, some of the more important actions up to late 2003 include the following:
  April 16, 2000 – Washington, DC, IMF
  May 1, 2000 – Global, May Day protests
  July 29, 2000 – Philadelphia, Republican National Convention
  August 11, 2000 – Los Angeles, USA, Democratic National Convention
  September 11, 2000 – Melbourne, World Economic Forum
  September 26, 2000 – Prague, Czech Republic, World Bank/IMF
  November 20, 2000 – Montreal, Quebec, G20 meeting
  January 20, 2001 – Washington, DC, Bush inauguration
  January 27, 2001 – Davos, Switzerland, World Economic Forum
  April 20, 2001 – Quebec City, Canada, Summit of the Americas (FTAA)
  June 15, 2001 – Gothenburg, Sweden EU Summit
  September 29, 2001 – Washington, DC, Anti-capitalist antiwar protests
  February 1, 2002 – New York City, USA / Porto Alegre, Brazil World Economic Forum / World Social Forum
  March 15, 2002 – Barcelona, Spain EU Summit
  April 20, 2002 – Washington, DC (War on Terrorism)
  November 4 to November 10 – Florence, Italy, First European Social Forum
  June 26, 2002 – Calgary, Alberta, and Ottawa, Ontario, G8 summit at Kananaskis, Alberta J26 G8 Protests
  September 27, 2002 – Washington, DC, IMF/World Bank
  Weekend of February 15, 2003– Global protests against war on Iraq, about 12 million antiwar protesters
  July 28, 2003 – Montreal, Quebec
  September 14, 2003 – Cancún, Mexico – Fifth Ministerial of the WTO collapses
  October, 2003 – regional WEF meeting in Dublin, European Competitiveness Summit, canceled
  November 20, 2003 – Miami Mobilzation against the Free Trade Area of the Americas (FTAA)
  From the Wikipedia article on "Anti-Globalization.": en.wikipedia.org/wiki/Anti-globalization_movement#Is_.22anti-globalization.22_a_misnomer.3F (June 15, 2005).

War), there are significant differences as well. To begin with, though most members of the Global Justice Movement would probably consider themselves left of center, its constituency is actually surprisingly politically diverse, with a sizable number of anarchists, on the one hand, and "natural" capitalists, on the other. One factor that might account for this greater diversity relative to the movement's counterparts in the sixties and seventies is the sudden and unforeseen collapse of Soviet communism in 1989. (While the dictatorships of Stalin and Mao had already served to disillusion many Western communist sympathizers, the persistence of a world communist movement centered in the USSR still offered the possibility of a tangible alternative to the dominant American capitalist model.) A second factor has to do with the more generally diverse character of the demographic base from which the movement draws its members. Though still attracting a large number of youth, many are now in the second half of life and, though they grew up in the sixties and seventies, a good number of these have exchanged their earlier single-minded idealism for a more nuanced, seasoned, or in any case harder-to-classify political stance.

Another significant difference is that the Global Justice Movement has revealed itself as an explicitly and self-consciously planetary phenomenon. It is true that the earlier counterculture, though focused in North America and Europe, was international in scope and global in impact. The relative simultaneity of mass protests (Berkeley, Paris, Mexico City, Beijing), however, seems to have been the expression of a spontaneous groundswell—what Jung would call a *synchronicity*—in the collective consciousness (especially among the youth), and not an intentionally coordinated international effort. Obviously, the biggest factor that accounts for this difference is the rise of the Internet in the intervening years. Never before was it possible to disseminate information so widely and quickly, or so efficiently to coordinate the actions of so many. A more general consideration has to do with the sharp rise in global economic interdependence and the growing consciousness of global threats (climate change, terrorism). Along with the antiwar demonstrations of February 15, 2003, to protest the invasion of Iraq, the global justice protests are to date the most impressive examples of Internet-facilitated mass rallies.

Finally, though the popular ecological and environmental* movements began in earnest in the sixties and early seventies (1962: Rachel Carson's *Silent Spring* is published; 1970: the first Earth Day is celebrated; the U.S. Environmental Protection Agency begins operations; 1971: Greenpeace is founded; 1972: the U.N. Environment Program is established; 1973: passing of the U.S. Endangered Species Act), and though the hippie contingent of the counterculture was particularly sensitive to environmental concerns, it is only in recent years that one begins to see a more systematic coordination of social justice and environmental activists. This is most apparent in the case of the movement for environmental justice, which highlights the way the poor and socially oppressed suffer most from the effects of environmental degradation. As far as the global justice/anti-globalization movement is concerned, however—and despite the explicit recognition in much of the literature of the inseparability of social justice and environmental issues—the dominant note is clearly on the social side of the relation (which is evident in the word *justice,* a term borrowed from the realm of social relations under the rule of law).

While it is to such groups as *Greenpeace* and *EarthFirst!* that one must turn for a more focused emphasis on environmental or ecological issues, there does seem to be a trend among activists in general toward more integrated approaches or perspectives on the

---

* I am aware of the distinctions often made, though I do not press them here, between the science(s) of ecology—which date as far back as the eighteenth century (Linnaeus, Gilbert White), and certainly the nineteenth century (Thoreau, Haeckel); the American conservation movement—which begins in the late nineteenth century (G. P. Marsh); and the wider environmental movement, which dates to the 1960s and 1970s (see D. Worster's *Nature's Economy: A History of Ecological Ideas.* New York: Cambridge University Press, 1994).

current planetary situation. A leading example is Bioneers, founded in 1990, which describes itself as "a nonprofit organization that promotes practical environmental solutions and innovative social strategies for restoring the Earth and communities."[122] These solutions and strategies are guided not only by the ideal of *sustainability* (conservation and the shift to renewable energy), but also by *restoration*, which addresses "the interdependent array of economics, jobs, ecologies, cultures, and communities." Toward the end of its Mission Statement, we read:

> INTERDEPENDENCE...TEACHES US THAT THERE ARE NO SINGLE ISSUES because it's one whole that can be addressed only by bringing together all the parts. Bioneers gathers people at the crossroads of ecological restoration, human health and social justice. There is only one cause—it is all of them.[123]

We will return to the themes of holism and interdependence in the next two chapters. Here I would simply note that Bioneers, with its yearly conferences and its spin-off radio broadcasts, affiliated programs, and publications, constitutes a leading creative force in the evolving movement toward a more integrative activism.

Another significant element in the Bioneers mission statement, one that distinguishes this organization from the majority of activist groups—whether oriented primarily to issues of social justice or to those of the environment/ecology—is its recognition in principle of the importance of "spirit" for a truly effective approach to our planetary ills. "Uniting nature, culture, and spirit," we are told, "Bioneers embody a change of heart—a spiritual connection with the living world that is grounded in social justice."[124] Still, judging from the most recent conference program, where only two out of approximately seventy presentations explicitly address spiritual concerns (of the remaining presentations, about two-thirds are environmentally/ecologically oriented, with about one-third oriented to social justice issues), nature is the primary concern, with culture second and spirit a somewhat distant third. Despite the reference to "priests and shamans" among the list of potential bioneers (which, we read, include "scientists and artists, gardeners and economists, activists and public servants, architects and ecologists, farmers and journalists, priests and shamans, policymakers and citizens"), one expects that the kind of spirituality toward which such an organization would be drawn would be strongly Earth-based, immanentist in orientation, and suspicious of (or perhaps simply uninterested in) what could tend to be perceived as "ungrounded" metaphysical claims.

Even in the absence of explicitly religious or spiritual talk, however, it is clear that, for such as the bioneers, as indeed for most environmental/ecological activists, the fate of the Earth, as an "ultimate concern," is itself the occasion for a most passionate form of spiritual engagement. At the heart of this engagement is the recognition of radical interdependence, or what I would call an irreducible *global solidarity*. (I shall have more to say about such interdependence and the cosmic dimensions of this solidarity in the final chapter.) The benefit of the word *solidarity* over *justice* here is that it connotes *both* the ideal of freedom or liberation from oppression *and* the participatory character of the human (and indeed earthly and cosmic) situation. It is this solidarity that has prompted Edgar Morin to propose the emergence of a third kind of religion that, perhaps, could satisfy the spiritual longings of humanity in the grip of planetary crisis. According to Morin, the first kind of religion, which began to be eroded from the time of the Enlightenment, was a religion of salvation, of an otherworldly God or gods. The second kind of religion, typified in both Marxism and positivism or scientism, did not recognize itself as a religion, though it still held up the promise of (this-worldly) salvation. The third kind of religion would be a "religion in the minimal sense [suggested in one derivation of the word:

from *re-ligare:* to join back together]"[125] and "would involve a rational undertaking: to save the planet, to civilize the Earth, to unify
humankind while safeguarding its diversity."[126] At the same time,
however, it would be

> a depth religion, uniting people in suffering and death. It would
> not promise any primary or ultimate truth.... Such a religion would
> lack any providence, any shining hereafter, but would bind us
> together as fellows in the unknown adventure.
>
>   Such a religion would not have promises but roots: roots in our
> cultures and civilizations, in planetary and human history; roots in
> life; roots in the stars that have forged the atoms of which we are
> made; roots in the cosmos where the particles were born and out of
> which our atoms were made....
>
>   Such a religion would involve a belief, like all religions but,
> unlike other religions that repress doubt through excessive zeal, it
> would make room for doubt within itself. It would look out onto
> the abyss.[127]

Morin's eloquent call for a new kind of religion will doubtless
seem godless to traditionalists and perhaps too "spiritual" to secular
humanists. In any case, it is no easy thing to look out onto this abyss.
There is nowhere for the eyes to rest, and the body flinches or even
recoils in fear at what the mind cannot grasp. At the same time, to
the extent that we can remain open to "the fundamentally irrational-
izable," to the "creative and originary [*génésique*] ground"[128] of the
cosmos that has birthed us, we make room for the actualization of
our deepest humanity. We make room for the further emergence of
those spiritual ultimates—mutual understanding, forgiveness, compassion, and love—which alone might kindle a steady light in the
looming darkness, and perhaps spare us the worst in the years ahead.
This abyss, Morin has said, "this breach (*brèche*) in the midst of our
knowledge is also a mouth (*bouche*) struggling to speak."[129] It would
speak to us of the pregnant silence that supports and surrounds our

every utterance, a silence that can both terrify and console. It would also, however, give voice to the billions of poor and oppressed, to the millions of species facing imminent extinction, and to the Earth itself in is long travail.

11

FOR THE SOUL OF THE WORLD

WHILE the impetus behind the more explicitly activist wing of the counterculture fed into the many groups devoted to social justice (such as the New Left) and environmentalism/ecology, the other, more theoretically and consciousness-oriented wing gave birth to a related but distinct front, consisting of the various exponents of what is often referred to as the New Paradigm. As was the case at the height of the counterculture, there is considerable cross-over between both fronts. This is clearly the case, for instance, with the Bioneers, who, despite the temperamental emphasis on practical solutions, are clearly sympathetic to the movement(s) considered in this chapter. Nevertheless, then as now, the distinction is real enough to warrant a separate treatment of the more consciousness-oriented front.

The various articulations of the New Paradigm must be seen in the more general context of the New Age movement. The notion of the "New Age" has its deepest historical roots in the apocalyptic or millennarian dimension[130] of the biblical Great Code, and therefore can be seen as a fundamental expression of the third moment of the fundamental pattern whose spiraling path has, as we have seen, both anticipated and in some sense created the conditions for the birth and transformation of the Planetary Era. One could point to a long list of figures for whom the intuition of a New Age, if not the term itself, plays a central informing role—from the author of Revelation and St. Paul, through certain medieval theologians (such as Joachim and his followers) and Renaissance esotericists, through Puritan sectarians, Enlightenment and Romantic/Idealist philosophers, the American founding fathers, and nineteenth and twentieth century socialist and utopian visionaries, to the theosophist Alice Bailey (who seems to have been the first to propose the term in its more or less contemporary sense) and such late-modern religious movements as the 1950s UFO cults. According to Hanegraaff, however, the New Age as a distinct *movement* "emerged in the second half of the 1970s, came to full development in the 1980s and is still with us at the time of writing."[131] The movement, in other words, seems to have taken off just as the counterculture of the sixties and early seventies was beginning to wane. "The New Age movement," writes Hanegraaff, "is commonly, and rightly, regarded as rooted in the so-called counterculture of the 1960s."

> Indeed, many of the concerns which can still be found in the movement of the 1980s and the early 1990s were present in the 1960s [and, as we have seen, at the turn of the twentieth century, at the height of the Romantic/Idealist movement, and during the Renaissance], including the expectation of a New Age—the "age of Aquarius" as celebrated in the musical *Hair*—which would replace the culture of the status quo.[132]

I must refer the interested reader to Hanegraaff's masterful study of the New Age movement for more details. In the remainder of this chapter, I will focus on two principal manifestations of the New Paradigm (a term that is preferred by many who might feel uncomfortable with the term *New Age*)—namely, the so-called *new science*, on the one hand, and *transpersonal psychology* (or more generally, transpersonal studies), on the other.

In his treatment of "new age science," Hanegraaff draws particular attention to the "holographic paradigm" associated with David Bohm and Karl Pribram, the paradigm of "self-organization" associated with the school of Prigogine, Rupert Sheldrake's hypothesis of "formative causation," and Lovelock's "Gaia hypothesis" (now recognized as "Gaia Theory."[133] All of these paradigms can be seen as expressions of Thompson's "Complex Dynamical Mentality," the emergence of which, as we have seen, he dates to the spiral turn around the threshold of the twentieth century, but that has even deeper roots in earlier turns of the spiral. The *holographic paradigm* refers to a general conception of the nature of reality—initially physical reality, but ultimately encompassing mind or consciousness as well—which stands as a radical alternative to the dominant Cartesian–Newtonian view. This view, in which reality is conceived mechanistically as consisting of essentially inert particles in merely external relation to one another and acted upon by external forces, is considered by representatives of the New Paradigm to be largely responsible for the widespread alienation and fragmentation of the late modern period.[134] The holographic paradigm, by contrast—which bases itself on a certain reading of quantum (and to an extent also relativistic) physics—conceives of reality, at least at its most fundamental level, as "undivided wholeness in flowing movement."[135] This wholeness is manifest at the physical level in the complementary relation between such classically opposed terms as *space* and *time, matter* and *energy, wave* (or *field*) and *particle, position* and *velocity* (or *momentum*), and perhaps even more so in the mysterious phenomenon of "entanglement," or non-locality (of which Einstein spoke derisively, before it was experimentally confirmed as "spooky action at a distance").[136] The phenomenon of non-locality points to the realization that, at a certain level of depth or subtlety, all of matter/energy participates in or as a unitary or integral reality—Bohm's "unbroken wholeness" or, more technically, what he describes as the "implicate order"—without the spatial or temporal divisions we normally take for granted. Awareness of this essential wholeness, it is believed, could play a critical role in establishing the kind of cosmic solidarity envisioned by the activists considered in the previous chapter. (I shall return to the notion of cosmic solidarity in the next chapter.)

Sheldrake's *theory of formative causation*[137]—also known as the theory of *morphogenetic fields or morphic resonance*—does for the life sciences what Bohm's theory of the implicate order does for physics. Here too, the properties of wholeness and non-locality are manifest, in this case with the emergence, maintenance, or transformation of organic forms—with embryogenesis, metabolism, or speciation, respectively. Such processes, in Sheldrake's view, cannot be reduced to the mechanical interaction of molecules. The reason is simply that, though we can say with some certainty that particular sequences of molecular or chemical interactions seem to accompany, and even stand in some causal relation to, these processes, there is nothing in the molecules or sequences themselves that literally represent or contain the forms in question—all we have are the equivalent of building instructions for the linking together of particles, or clusters of particles. The relative constancy of the forms—whether it be the bilateral symmetry of the body or the number and arrangement of fingers and toes, for instance—suggests the presence or influence of a field. A favored analogy here is that of a television, where no one with a knowledge of the technology would assume, because the image on the screen is dependent on the circuitry inside the box, and because the image changes through manipulation of the hardware, that the circuitry or hardware is therefore the ultimate cause of the forms one sees on the screen. In both cases—with the television and with organisms—the primary cause of the forms is information carried by an invisible field that is not precisely localizable. In the case of so-called M-fields, moreover, it is hypothesized that there is no diminution of information or formative potential over space or time, which is to say that there is genuine non-locality.

Sheldrake has generalized the principle of formative causation to include all relative constancy of form, whether organic, inorganic (as in crystals), behavioral (not only with instincts, for instance, but with learned behavior), or perceptual-ideational. In a similar fashion, Bohm has extended his theory of the implicate order beyond the realm of the quantum to the realm of the mind or consciousness, both individual and collective. For both theoreticians, as for those who are inspired by them, the properties of wholeness and non-locality are valued not only for the insight they provide into the nature of isolated physical, biological, or psycho-social phenomena, but for the suggestion of an alternative to the reductive and fragmenting qualities of the dominant Cartesian–Newtonian paradigm.

A third theory is that of *self-organization*. Also concerned with the emergence, maintenance, and mutation of forms, but proceeding from cybernetics, systems theory, and Prigoginian thermodynamics, the theory of self-organization focuses on what could be called a kind of dynamic holism—on patterns and processes instead of on substances in mechanical interaction, on "feedback" or circular, rather than simple linear, causality. While some have warned of the potentially reductive character of some systems thinking—whether in a kind of simplifying or totalitarian holism,[138] or in a certain blindness to interiority[139]—the theory of self-organization not only provides a description of the nature of evolutionary or developmental processes, but also, as with the theories of Bohm and Sheldrake, suggests the more general possibility of overcoming some of the deficiencies of standard scientific materialism. Along with the notion of dynamic holism, the theory makes room for the emergence of genuine novelty (with the idea of "order from chaos"), for the importance of global transformations ("phase shifts" or "bifurcation points"), a certain unpredictability of evolutionary outcomes, and the potentially critical effect of very small actions or changes in the organization of complex systems (the co-called butterfly effect).

Finally, there is Lovelock's *Gaia theory*,[140] which might be considered a special application of the theory of self-organization. While researching ways of determining the possible existence of life on Mars, and inspired by imagining the Earth as seen from space, Lovelock came to the sudden realization that the Earth could be conceived as a single living organism. He and Lynn Margulis later brought their combined knowledge of biology and earth sciences to bear on several self-regulatory processes (for instance, the role of bacteria in the relation of atmospheric temperature to the sequestration of carbon dioxide) to demonstrate that the Earth does indeed manifest the qualities of a living organism as defined by such systems thinkers as Varela, Maturana, and Morin—namely, feedback loops or circular causality in the interests of self-organization.

While Lovelock has denied that his theory includes the claim that the self-organizing behavior of the Earth is purposive—and by implication, that the Earth is sentient in the manner that other organisms are, or that it (or she) possesses conscious intelligence or even a soul (the *anima mundi*)—it is clearly in this direction that the theory has been taken up by most New Paradigm or New Age writers. Obviously, calling the Earth by the name of the ancient Greek Earth Goddess is not calculated to encourage restraint in such matters. In any case, and though less self-consciously than Bohm or Sheldrake, Gaia theory has become an integral part of what Hanegraaff rightly considers a coherent, if still developing, *Naturphilosophie*—a comprehensive philosophy of nature that seeks not only to synthesize elements from leading-edge science, but also to set the scientific understanding of the cosmos in a wider context that recognizes the irreducibility and centrality of mind, consciousness, and spirit.

Hanegraaff's association of the new science with the earlier Romantic and Idealist *Naturphilosophie* is apt, for, as we have seen (in chapter 6), it is in this earlier iteration of the third moment that we have the first articulations of both nature or the cosmos and history as an evolutionary, self-organizing totality (generally called the Absolute, in the sense of the Whole). Most representatives of the

new science are unaware of the degree to which their core insights—
including, for instance, a variant of the Gaia hypothesis—had already
been anticipated almost two centuries earlier in the philosophies of
Schelling and Hegel. Sadly, as Antoine Faivre has remarked, "our
own traditions, even in these areas, are generally unknown." [141] *

The great wave of speculative thought and system building of
the German Idealists followed the shock of the French Revolution
and continued through the rise and fall of the Napoleonic Empire.
Similarly, the rise of the New Paradigm almost two centuries later
followed the shock of the Vietnam War and continued through
the emergence of the United States as the dominant superpower
following the Cold War. Also, just as the Romantic and Idealist
Naturphilosophie was accompanied by a philosophy of spirit
(*Geisteswissenschaft*), so the new science was paralleled by the emergence of *transpersonal psychology*.

While we can look as far back as the origins of experimental psychology with G. T. Fechner (who was a student of Lorenz Oken, himself a disciple of Schelling), with William James, and certainly with
Jung (who explicitly placed himself in a line of thought running from
Kant and Goethe, through the Romantics, to the birth of psychoanalysis), the formal beginnings of transpersonal psychology date to
1969 to 1970, when the movement was launched by Anthony Sutich,
Abraham Maslow, and Stanislav Grof. This was also the year that the
Journal of Transpersonal Psychology began publication. [142] It was Grof
who proposed the term *transpersonal* to describe the "fourth force"
that would succeed humanistic psychology as the "third force" (the
first two forces being psychoanalysis and behaviorism, respectively).
Grof was also the first to propose a new, comprehensive "cartography

---

* This is perhaps especially true for the American collective psyche, given its dominant position as the frontier of the current second general moment of differentiation (leading to dissociation and disenchantment, and therefore to a generalized
amnesia with respect to earlier countercultural traditions).

of the psyche," which, in its mature formulation, would not only integrate the perspectives of Freud, Adler, and Jung, but would draw significantly from both the leading developments in the new science and the world's spiritual and mystical traditions. (The first, long chapter of his book *Beyond the Brain*, [143] "The Nature of Reality: Dawning of a New Paradigm," is still one of the best summaries of both wings of the New Paradigm.)

There are three central concepts in Grof's model. The first is that of the *transpersonal*, which refers to a class of experiences of "mind-at-large" that involve the transcendence of the spatial and temporal boundaries that normally define the ego or separate-self sense. These experiences can range from encounters or identification with other beings (both human and nonhuman) all the way to mystical union and dissolution into what Grof describes as the "supracosmic and metacosmic void." [144] However "nonordinary" in the context of modern secular society and either denigrated or ignored by adherents of the Cartesian–Newtonian paradigm, such experiences have been well documented for millennia by the world's spiritual traditions. While they continue to occur spontaneously and are even cultivated by small numbers of people, Grof was able to experience them firsthand and record them by the hundreds in the context of his pioneering work with LSD psychotherapy in Prague and the Maryland Psychiatric Institute. Though it arose within the context of modern medical and psychoanalytic therapy, Grof's research was naturally taken up by the wider xounterculture in which, of course, the unstructured and unsupervised widespread use of LSD was to play a catalyzing role. [145]

The second central concept in Grof's model is that of the *perinatal*. Though initially an expansion and deepening of the Rankian insight into the effect of the birth trauma on the development of personality, the concept eventually unfolded in the direction of a more complex and nuanced understanding of the death/rebirth archetype, where the life-and-death struggle of the fetus at birth is seen as the prototypical encounter with the universal, or archetypal, process of deep transformation. The archetypal character of this process—which in this case has to do with the archetype of process or transformation itself[146]—is revealed in its triphasic deep structure. As I alluded to in chapter 2, this process mirrors the triphasic structure of the Hegelian dialectic. The movement from identity, through difference, to a new identity (which Grof speaks of in terms of the four perinatal matrices) is manifest in the (developmentally critical) passage from conception to birth, in all major life transitions, and in all experiences that involve an ego death and an opening into the transpersonal realms.

The third central concept in Grof's model is that of the *holotropic,* which has obvious resonances with the more general notion of the holographic paradigm. The term means, literally, "turning or moving toward wholeness," and refers in the first place to the observation that the psyche has an innate drive to heal its (constitutional or developmentally incurred) lesions and to actualize its deeper (and especially transpersonal) potentials. In this sense the term represents an amplification of Jung's understanding of "individuation" as the path of wholeness, with less emphasis on the actualization of individualized selfhood. At the same time, however, the resonance with the concept of the holographic suggests that these deeper potentials involve the expansion of selfhood to include potential identification with all aspects of the cosmos, and indeed, at the limit, with the "supra- and meta-cosmic void." Here the concept of the holotropic merges with that of the transpersonal, and Grof's initially psychological model reveals its openness to a more explicitly mystical/metaphysical worldview.

The second major figure in the transpersonal movement to produce a comprehensive model—or to be more precise, at least three successive models—is Ken Wilber. From the beginning, with his call for a *psychologia perennis*—which is to say, a psychology that seeks to articulate itself in the context of the general outlines of the so-called perennial philosophy—Wilber has openly embraced the explicitly spiritual/metaphysical dimensions or levels of reality. He did so initially with reference to the "full-spectrum" analogy and then with the *"integral"* idea of taking into consideration *all levels, all quadrants."* By this he means the pre-personal, the personal, and the transpersonal (levels), and the crossing of two axes: interior/exterior and individual/ collective (quadrants). Though critical of the claim among many New Age enthusiasts that the paradoxes of quantum theory can legitimize the worldview of traditional mystics,[147] he nevertheless saw his work as directed to the working out of the emerging "new paradigm," in the sense of "an *overall* knowledge quest that would include not only the 'hard ware' of physical sciences but also the 'soft ware' of philosophy and psychology and the 'transcendental ware' of mystical-spiritual religion."[148]

While he does not claim to have made a serious study of Hegel, Wilber does admit, in the original version of *Up from Eden*, that "the shadow of Hegel falls on every page of this book."[149]* The element of Hegel's thought that seems to have had such a strong impact on his theorizing at this point is the crucial distinction between the first and the third moments of the fundamental pattern—the difference, that is, between the relative simplicity of the origin of development (which contains the whole as mere potential) versus the complexity of the goal (where potentials have been actualized). It is the honoring of this distinction that led him to reject the first phase of his thought ("Wilber I," which was more or less Jungian) as flawed by the so-called *pre/trans fallacy* (in this case, confusing the pre-personal with the transpersonal) and, as he would later characterize it, as being an expression of "retro-Romanticism."

Though, to my mind, mistaken in his assessment of both Jung and Romanticism, the distinction in question is critical. While Wilber provides many insights into the dangers of conflating the two moments, it would be more accurate to consider those elements of

---

* The line, for some reason, was deleted from subsequent editions.

Romanticism that Wilber denigrates as "retro"—i.e., nostalgia for the participatory "innocence" of childhood, heartfelt identification with Gaia, emphasis on the imaginal, etc.—as representing the re-emergence, in Western consciousness, of the "magical" and "mythic" structures that had been dissociated through the triumph of the mental egoic culture of the Enlightenment. As such, and to the extent that these were not absolutized but taken up into a new, more integrated consciousness, Romanticism actually signals the emergence of a truly "integral" (Wilber's "centauric") consciousness.*

What Wilber also seems to miss is the fact that the magical and mythical worldviews that the Romantics sought to rehabilitate cannot be isolated from the transpersonal elements in which they were embedded, so that what was being revived was not simply past, "pre-personal" structures, but the complex and *integrated* type of world-views where the transpersonal is neither split off from the personal nor denied altogether (both of which options represent the historical endpoints of rational-egoic culture). In other words, Wilber's critique of the Romantics as "regressive" only holds if one takes Romantic discourse/rhetoric *literally*. To do so, however, is profoundly *un*romantic, since of central importance to the Romantic spirit is the revaluation of imagination, metaphor, and symbol. "Unless ye become *as* children" is not a call to infantilism. The "as" points to childhood as an ana-logue of the "higher" integration to which the Romantics aspire. It is a matter, moreover, of *becoming* as children, which means embracing the process of conscious/conscientious striving to actualize what one already potentially is. "The typical Romantic ideal," as M. H. Abrams says, "far from being a mode of cultural primitivism, is an ideal of strenuous effort along the hard road of culture and civilization."[150] To the extent that some Romantics (or contemporary New Paradigm thinkers) fall prey to the deadening literalism they otherwise abhor, then of course Wilber's critique would be justified. Such "Romantics,"

---

* I make the same argument for individual development in Kelly, 1998.

we might say, are the most naive expressions of the type, if indeed they can legitimately be considered as belonging to the type at all.[151]

It is in continuity with the tradition of such an integral Romanticism that one must consider the archetypal-astrological perspective of Richard Tarnas. Tarnas has rightfully gained widespread international respect for his *Passion of the Western Mind*, a careful, deeply insightful, and elegant narrative of the evolution of Western thought. The underlying hermeneutic of the narrative draws especially from the thought of Thomas Kuhn, Hegel, Jung, and Grof, arguing that the movement from the Greek and Christian, through the modern, to the postmodern worldviews can be understood as an organic and dialectical unfolding of a collective mind or psyche. What most readers of the *Passion* did not realize, however—a realization that would have to await the publication of his next book, *Cosmos and Psyche: Intimations of a New World View*—is that Tarnas had already come to the conviction that the mind or psyche in question is not to be conceived *merely* hermeneutically in the manner of an "as if," but rather in more explicitly metaphysical terms as what the ancients, the Renaissance magi, and the Romantic and Idealist philosophers called the *anima mundi,* or World Soul. The main difference between Tarnas's view of the World Soul and that of his predecessors, however, is, to begin with, that Tarnas has arrived at his conviction through the accumulation of a staggering wealth of empirical observations, a summary of which is presented in *Cosmos and Psyche*. Second, Tarnas's view is informed by the leading developments in (quantum) physics, (postmodern) philosophy, and (Jungian/archetypal) psychology that have marked the development of Western thought over the last hundred or so years.

I cannot do justice to Tarnas's argument here (it is engaged more fully in appendix II). Its core, however, rests on the following claim: the major events of Western cultural history—from wars and revolutions to major scientific discoveries and the birth and ongoing evolution of influential social movements (such as the push for civil rights or the expression of religious renewal)—are consistently and meaningfully correlated with the observed angular positions of the planets (especially the three—Uranus, Neptune, and Pluto—discovered by telescope in the modern era). To give just one out of countless examples (which I choose because of its obvious relevance to the movements I have focused on in this study):

> An especially notable planetary alignment in Western cultural history that involved the 120° trine aspect was the rare "grand trine" configuration of Uranus, Neptune, and Pluto that took place between 1766 and 1777.... Such a grand trine involving the three outermost planets occurred only once in the modern era. The period of that alignment coincided with the very height of the Enlightenment, when there took place many of that era's most distinctive milestones such as the completion, by Diderot and the other *philosophes*, of the *Encyclopédie*, the eighteenth century's great bible of intellectual emancipation; the writing of the Declaration of Independence...; and the beginning of the American Revolution as a self-conscious expression of Enlightenment ideals and principles....
>
> This same period of the later 1760s and 1770s also coincided with the great birth of Romanticism in Germany, introducing that seminal and profound cultural impulse into the European mind. From the work of Herder and Goethe during these years emerged a new conception of nature, spirit, and history—and of language and art, intellect and feeling, interiority and imagination, sensuality and spirituality, humanity and divinity—that would dramatically bear fruit...during the immediately following Uranus–Pluto and Uranus–Neptune axial alignments from the 1790s through the 1820s (and indeed beyond those periods to the most recent such alignments of the 1960s and 1990s). In addition, virtually the entire central generation of Romantics was born during the decade of this Uranus–Neptune–Pluto grand trine: Wordsworth, Coleridge, Schelling, de Staël, the Schlegel brothers, Schleiermacher, Hölderlin, Novalis.

The powerful confluence of brilliant creativity and the urge for freedom and change (Uranus), of imagination, spiritual aspiration, and charismatic idealism (Neptune), and of nature, evolution, instinct, and eros (Pluto) that began to be brought forth into the world at this time and was then given poetic and philosophical form by the generation born during this period corresponds exactly to the character of a grand trine involving these planets and archetypal principles. Remarkably, during the period when the three planets were in especially close alignment, in 1769 to 1770, three world-historic individuals were born whose lives and influence especially embodied this archetypal confluence: Napoleon, who was born with *Mars* [archetypally associated with war] on the grand trine; Beethoven, who was born with *Venus* [associated with art] on the grand trine; and Hegel, who was born with *Mercury* [associated with the mind] on the grand trine.[152]

The cogency of this particular set of "meaningful coincidences"—this is how Jung defined his notion of synchronicity (which Tarnas adopts and elaborates)—is considerably amplified by the preceding hundreds of pages of careful argument and many dozens of similarly (to the Cartesian–Newtonian mindset, at least) "improbable" correlations. The full force of the evidence, however, is available only to those who actually take the time and effort to look through the archetypal-astrological telescope themselves. Those who have neither read the book nor learned enough practical astrology to track the correlations between planetary alignments and the significant events of their inner or outer lives will need to suspend judgment and, for the moment at least, be open in principle to the profound implications of Tarnas's research. I have already drawn attention to what I see as the main implication—namely, compelling evidence for the existence of the World Soul, whose deep structures or patterns (the archetypes) are the common ground of intelligibility of both nature and spirit, cosmos and psyche—which of course was a guiding insight of Renaissance esoteric and Romantic/Idealist speculation. That we

have reason (and imagination!) to believe that we participate in such a
World Soul means that the cosmos is once again enchanted (or, more
precisely, that we are reawakening to its inherent enchantedness), that
the natural world and human consciousness, the individual and the
collectivity, the human and the divine, are distinguishable though
not separable elements or "moments" (as Hegel would say) of an integral totality or living wholeness.

While Tarnas warns of the dangers of using the archetypal-astrological perspective for predicting future events—since the planetary
alignments indicate at most the main features of the archetypal landscape that lies ahead and do not say anything about concrete particu-
lars—the perspective does suggest something in the way of timing for
the next turn of the spiral we have been tracing. According to Tarnas:
"If we can judge by past experience, the most significant and potentially dramatic configuration on the horizon is the convergence of
three planetary cycles that will produce a close T-square alignment of
Saturn, Uranus, and *Pluto* during the period 2008 to 2011." [153] Though
we will probably only really know in retrospect, this four-year period
could very well coincide with the beginnings of a new third moment
or phase, which in this case one might expect would correspond to a
more intensified or generalized manifestation of the spirit of the New
Paradigm in wider spheres of culture and society. From what we have
seen, we can anticipate that this new phase would resonate deeply
with prior third phases, and especially (given the planets involved)
with that of the sixties counterculture. The "already approaching Uranus–Pluto square alignment that will extend through 2020,"
writes Tarnas, "points to the possibility of a significant cyclical
development of the cultural impulses and archetypal dynamics that
emerged during the 1960s." [154] When the trajectory of the tightening
spiral is combined with the archetypal-astrological perspective, therefore, we have grounds to expect that the next major countercultural
impulse—if it is able to overcome the hardening forces of reaction,

on the one hand, and navigate the generalized anarchy or chaos that appears ever more likely, on the other—could span the period 2008 to 2020.* As Tarnas notes, the "coming to power of the generation born during the Uranus–Pluto conjunction of the 1960s and its aftermath will certainly play a crucial role." [155] We shall have to wait and see.

---

* The epochal victory of Barack Obama was significantly carried by an Internet-facilitated campaign that found the support of the majority of progressives, women, youth, and nonwhites. At the time of writing, it remains to be seen to what extent this administration will mostly satisfy or disappoint the more countercultural contingent that rallied to support him.

12

FIRST LIGHT
TOWARD A PLANETARY WISDOM CULTURE

THE evolutionary spiral we have been tracing, especially since the birth of the Planetary Era, has involved not only the differentiation, but also the dissociation between such pairs of elements as the individual and the collective, humanity and nature or the cosmos, and the sacred and the profane. The modern West, and therefore increasingly the rest of the planet, has tended in the direction where one of the elements of each pair is dominant or has sought to become so: the human has eclipsed the natural through the construction and proliferation of artificial environments and through the subjugation, and now the devastation, of the biosphere. A one-sided, atomistic individualism has helped generate a reigning culture of narcissism, which in addition is correspondingly vulnerable to the phenomenon of mass-mindedness.[156] The systematic disenchantment of the world has gone hand in hand with the growth of increasingly virulent forms of religious fundamentalism.

At the same time, and thankfully, there continue to be movements of resistance and countercultural impulses throughout these dark times. And as we have just seen, there are indications—like the first light before dawn—that these movements and impulses might be about to coalesce into a new, creative, countercultural force with potentially world-transforming consequences. I am reminded here, however, of Jung's encounter with Ochwiay Biano, the Pueblo elder who shared his conviction that "with our religion we daily help our father [the Sun] to go across the sky.... If we were to cease practicing our religion, in ten years the Sun would no longer rise. Then it would be night forever." [157] We, too, who are sensitive enough to behold this first light, must be diligent in the performance of our "religion," which, as Jung at one point defines the term (following the formulation of Rudoph Otto), involves the "careful and scrupulous observation of...the *numinosum*." [158] We can take the concept of the "numinous" here to refer to the deeper promptings of the World Soul, to those symbols, values, and experiences that express our "ultimate concern" and speak to the realization that we are responsive to, and responsible for, if not the fate of the Sun, then at least the fate of the Earth community as it struggles to cross the threshold into a more authentically Planetary Era.

For those less comfortable with the concept of religion, even as it is understood by Jung, it will be readily agreed that we must seek to cultivate whatever habits of mind—and of heart and spirit—that might constitute the beginnings, at least, of a potentially transformative, because genuinely planetary, *wisdom*. As is the case with any deep and transformative work at the personal level, it is first of all necessary that we face the planetary *shadow,* which is to say that we acknowledge the continuing and in some ways unparalleled suffering on a global scale associated with the Iron Age, as Morin calls it, of the Planetary Era—the suffering of the still-growing numbers of poor and oppressed peoples; the suffering of the nonhuman members of the Earth community, so many species of whom have already been lost to the current mass extinction. To acknowledge this shadow is to *own* it, which does not, however, mean engaging in fruitless self-condemnation. Rather, it means accepting that, whatever our life circumstance, we are creative-destructive participants in the global drama, the course of which both determines, and is determined by, even the most seemingly inconsequential of our actions, thoughts, and feelings—especially when these are secretly guided or informed by root assumptions of the still-dominant paradigm. For instance, however much many of us might revile the policies of certain administrations, the subtle (or perhaps not so subtle) way in which people (myself included) sometimes demonize its leaders by referring to them by any number of readily available demeaning epithets only serves to reinforce the general climate of denial and dissociation.* Even in our opposition to the forces (corporate greed, nationalistic hubris, rampant consumerism, and so forth) that are creating ever greater numbers of poor and oppressed and are bringing the biosphere ever closer to chaos or collapse, all of us are complicit to varying degrees. (Who among us, for instance, does not depend upon, or benefit from, the petrochemical industry?) In this sense we *are* those we oppose.

To begin owning the planetary shadow in this way is also to make room in consciousness for potentially overwhelming levels of anxiety, grief, and despair. As Macy argues, however,[159] these feelings are already there, but are repressed or split off, not only because of the obvious threat they pose to an otherwise secure (because largely unconscious) ego, but precisely because of the false perception that, in confronting this shadow, we are faced with something radically *other*. Not only do we forget that it is we ourselves who cast the shadow—we also tend not to perceive the degree to which the pain and suffering we would alleviate comes to us from within as much as from without, from the deeper and subtler reaches of our true selves, which, as we shall consider below, we share with all members of the Earth community.

With this recognition of our participation in the planetary shadow as a kind of psychotherapeutic safeguard, we can proceed

---

* This is not to say, as a friend commented to me upon reading this passage, that a spade should not be called a spade. The metaphysical and theological problem of the nature of evil is of course irremediably vexed. What I am pointing to is the need for discernment in the experience of evil, where projection so easily occurs. For a profound and nuanced treatment of the problem of evil in the context of a planetary ethics, see Morin, 2004.

with greater confidence to the proposal of concepts or principles that might serve as the basis for a genuinely planetary wisdom. We have already encountered several candidates for such concepts and principles in our consideration of the successive inflections of the third moment, most notably in the various concepts associated with the New Paradigm. If we were to choose a single concept as the most fundamental, in that it is in some way presupposed or implied by all the others—recall our discussion of the notions of the holographic, morphogenetic fields, Gaia theory, the holotropic, the integral, synchronicity, and the World Soul—it would be the concept of *wholeness*. The word suggests not only the sense of totality, integrality/integrity, harmony, completeness, fullness, comprehensiveness, actualization, and realization, but also—through its deep resonance with its cognates, *healthy* and *holy*—the qualities of vitality, adaptiveness, sustainability, and creativity, on the one hand, and those of numinosity or meaning, purpose, wonder, and mystery, on the other.

While the terms *holism* and *holistic* have figured prominently in New Paradigm thought—precisely because of the above associations— one must be wary of the tendency to lapse into a subtler version of the kind of reductionistic thinking from which the New Paradigm is supposed to liberate us. The tendency is apparent in the common practice of presenting summary tables of the Old "versus" the New Paradigm, with such descriptors as "reductionistic," "mechanistic," "analytic," and "competitive" on one side, and "holistic," "organic," "synthetic," and "cooperative" on the other. As a long line of sages from Heraclitus, Lao Tzu, and Nagarjuna to Cusa, Bruno, Boehme, Blake, Schelling, Hegel, and Jung have seen, the wholeness in question is misconstrued when it is seen as heading a series of qualities set off against their opposites—which is why Jung followed Cusa in describing the Self, or the archetype of wholeness, as a *coniunctio* or *complexio oppositorum* (a "marriage" or "weaving together of opposites"). The "New," in other words, must embrace or integrate the essence and virtues of the

"Old," otherwise it remains partial and incomplete, and therefore *not* whole. This is not to say that the New Paradigm, or a truly planetary wisdom, must embrace or carry on the simplification, fragmentation, and pathological splitting or dissociation associated with old style reductionism. Still, there must be a place for methodical isolation and analysis, for careful definition and discrimination—for the moment of *difference*, in other words.*

In more concrete terms, a truly planetary wisdom must recognize the nobility as well as the shadow of the Old Paradigm, or to borrow Wilber's helpful distinction, it must honor its "dignity" along with its "disaster."[160] If we think in terms of the three phases of the larger arc—the premodern (mythic, and in this case, biblical), the modern (secular-scientific), and the planetary—it is clear that any proposal for a planetary wisdom must honor and integrate the hard-won accomplishments of modernity. Corresponding to the general moment of difference, these include the sociopolitical ideal of freedom (however difficult in practice it might be to implement, and even to define in unambiguous terms), as well as the cumulative (if partial and sometimes misleading) revelation of the nature of the cosmos through science and technology. A planetary wisdom, therefore, will not involve a wholesale repudiation of modernity and a literal return to origins, but rather (recall Hegel's notion of *aufheben*) a simultaneous fostering of the virtues and a rejection of the vices associated with the modern era. As we saw in the last chapter, there *is* a need for the recovery of a certain quality of wholeness that was more in evidence up until the triumph of modernity, a wholeness reflected in a more organic relation to an enchanted cosmos. But this earlier wholeness can never, nor should it, be literally recovered.

---

* In this connection, see Thompson (1986): "I hope that we will...through the cultural integration brought on by both the electronic technologies and the ecologies of Mind...come up with something more like a planetary cultural ecology in which difference is vital as the information that spells transformation" (92).

The goal of a new organicism and a re-enchanted cosmos can only be reached through the forward path of continuing individuation, as Jung would say, and not through regression to some less actualized and differentiated state. The "way back to Arcadia is closed forever," says Schiller, "onward toward Elysium!" And as Coleridge so nicely puts it: to "distinguish without dividing" is the only way to "prepare for the intellectual re-union of the all in one." [161]

It is with this idea of honoring difference/differentiation as an essential moment of genuine wholeness that I have proposed[162] the notion of *complex holism,* which to my mind could serve as an overarching concept or meta-principle in our search for a planetary wisdom. This principle—whose structure and dynamics correspond in obvious ways to the fundamental pattern (of movement from identity, through difference, to a new identity) in the evolution of the Western, and now planetary, worldviews—takes inspiration not only from Hegel's logic of the Absolute and Jung's formulation of the Self as *complexio oppositorum,* but also from Morin's wise reflections on the paradigm of complexity.[163] In particular, I have drawn on his understanding of three related principles: the dialogic, recursivity, and the holographic principle.

Morin defines the principle of dialogic as "the association of two [or more] principles which are at once complementary, concurrent, and antagonistic." [164] The most well known symbol of such an association is perhaps the yin/yang (or *taiji*) diagram. More concrete examples include the wave/particle duality in quantum physics, the nature/nurture relation (for example, with the influence of genes versus the environment in matters of health and illness), and the relation between the brain and mind or consciousness (more on this below). The quality of antagonism in the dialogic stresses the fact that the principles associated retain their relative autonomy and therefore resist being reduced one to the other, or being resolved in some final "synthesis."

The principle of recursivity describes the circular character of the process linking the associated terms. A process is recursive, writes Morin, when its end (or product) "nourishes the beginning, through a return of the final state of the circuit to and in the initial state." More particularly, a process is recursive wherein "an active organization produces the elements or effects necessary for its proper generation or existence." [165] We see this in all forms of living organization (from single cells to complex organisms, societies, and ecosystems)—for example, in the oxygen/carbon dioxide cycle where plants produce oxygen, which is consumed by animals, which produce carbon dioxide, which is consumed by plants. At a more abstract level, there is the causal relation between the brain and the mind: from one point of view, the brain produces or affects the mind (which we see in the case of brain damage or with the effects of certain chemicals introduced into the brain), while from another point of view (with the placebo effect, for instance, or with the relaxation response), such mental events as intention or imagination have significant effects on the structure of the brain. The recursive relation here can be expressed by saying that the brain produces the mind that produces the brain.

Finally, the hologrammatic principle holds that "the whole is in a certain way included (engrammed) in the part that is included in the whole." [166] We have already considered the nature of the holographic paradigm associated especially with the figure of David Bohm and his influential theory of the implicate order. The hologrammatic (or holographic) principle as Morin understands it is enriched by its relation to the principles of the dialogic and recursivity, which helps avoid a simplistic reduction to the whole.

The notion of complex holism, which I am proposing as a metaprinciple that might inform an emergent planetary wisdom, finds expression in the following *four planetary ideals:* cosmic solidarity, human unity, radical interdependence, and spiritual liberation. At the same time, and appropriately, consideration of these ideals helps reveal the deeper character of the principle of complex holism.

1) Cosmic Solidarity

While everything in the Newtonian cosmos is held together by the action of universal laws, these laws, like the forces with which they are associated, were initially thought of as imposed from without by the Deist creator upon an essentially inert collection of merely externally related particles in mechanical interaction. As we saw in the previous chapter, this mechanistic view of the cosmos leads to and legitimizes a generalized alienation and fragmentation. In the new cosmology, by contrast, the laws of nature are recognized as an integral expression of the evolving cosmos itself, which, according to the standard theory, is conceived as having emerged from the "Singularity" with the "Big Bang" (or the "primal flaring forth," to use Brian Swimme's more evocative phrase) some fifteen billion years ago. The complex holistic nature of this evolving cosmos, and our solidarity with it, is beautifully expressed by Swimme in his reflections on the "omnicentric universe." Though more commonly associated with Bruno, this insight into the omnicentric universe actually first appears with the early Renaissance mystical theologian Nicolas of Cusa.* For Swimme, however, the insight arose in response to two apparently mutually exclusive cosmological observations: first, that light from the origins of the universe reaches us from fifteen billion

---

* "Since it always appears to every observer," writes Cusa, "whether on the Earth, the Sun, or another star, that one is, as if, at an immovable center of things and that all else is being moved, one will always select different poles in relation to oneself, whether one is on the Sun, the Earth, the Moon, Mars, and so forth. Therefore, the world machine will have, one might say, its center everywhere and its circumference nowhere, for its circumference and center is God, who is everywhere and nowhere" (Cusa, 161). Cusa may have been inspired by the medieval aphorism describing the nature of God as a *"sphaera cuius centrum ubique, circumferentia nullibi,"* which, according to Koyré, "appears for the first time in this form in the pseudo-Hermetic Book of the XXIV philosophers, an anonymous compilation of the XIIth century" (Koyré, 1957, 279).

miles away (and years ago)—which implies that we are at the periphery; and second, the universe is expanding away from us in all directions—we are at the center. Thus, writes Swimme, the "central archetypal pattern for understanding the nature of the universe's birth and development is omnicentricity."

> The large-scale structure of the universe is qualitatively more complex either than the geocentric picture of medieval cultures or the fixed Newtonian space of modern culture. For we have discovered an omnicentric evolutionary universe, a developing reality that, from the beginning, is centered upon itself at each place of its existence. In this universe of ours to be in existence is to be at the cosmic center of the complexifying whole....
>
> To enter the omnicentric unfolding universe is to taste the joy of radical relational mutuality. For we know this body of ours could have been a giant sequoia....
>
> The consciousness that learns it is at the origin point of the universe is itself an origin of the universe. The awareness that bubbles up each moment that we identify as ourselves is rooted in the originating activity of the universe. We are all of us arising together at the center of the cosmos.[167]

From an evolutionary point of view, it is because all of matter/energy (and therefore also space/time) was "contained" in the original Singularity that there is such a thing as universal laws (or "habits of nature," as Sheldrake would say)—that there is a "cosmos," in the sense of a systematically and esthetically ordered whole, in the first place. This original Singularity would seem as well to have something to do with the phenomenon of non-locality considered in the previous chapter. Again, however, we are faced with the complex character of this cosmos, which is at once singular—in origin, and in some sense also for the non-local implicate order or "quantum vacuum," which Swimme calls the "all-nourishing abyss"—and infinitely multiple at the level of the explicate order or ordinary, consensual reality. The cosmos is *unitas multiplex,* to use one of Morin's

favorite expressions—or a "multeity in unity," as Coleridge puts it—
the features of which can only be expressed in a series of dialogically,
recursively, and holographically related terms. Along with space/time
and matter/energy, there are also, as we have seen, the fundamental
particle/wave and position/velocity relations. And as both relativity
and quantum physics demonstrate, each in their own way, we have
to include as well the relation between subject and object or observer
and observed.* In other words, it is not only that the physical "stuff"
of which we are made is inextricably woven into the complex fabric of
the entire cosmos, it is that this very stuff, which in some mysterious
way gives rise to or at least facilitates the emergence of our experience, is also (complexly) co-constituted by our experience of it. Our
solidarity with the cosmos is, at the most fundamental physical level,
thoroughly and complexly *participatory:* we participate as members
of the cosmic whole or totality, *and* we participate in bringing this
whole into manifestation or actualization. This is true not only in formal physical experiments involving measurement, but in every act of
perception, in even the slightest or subtlest of our interactions with
the surrounding world.[168]

For instance, quantum field theory suggests that the light from
the full moon hovering over the horizon, immediately prior to its
contact with your retina, is in some sense smeared across the whole
sky. This has to do with the wave character of light, the mathematical formalism for which (the wave function) describes the varying
intensities of the electromagnetic field, to which correspond different probabilities for finding a photon at a given coordinate in space/

---

* For Relativity, the interdependence of observer and observed is most apparent
with the notion of the *relativity of simultaneity,* where an event that is "past" from
one perspective (that is, from one inertial frame of reference) can be "future"
from another. In Quantum Theory, this interdependence is tied to the so-called
uncertainty relations (between particle/wave, position/velocity, and so on) and the
associated *uncertainty or indeterminacy principle.* In most interpretations, the relative weighting of either term in whatever pair under consideration is determined by
the choice of the conscious observer/subject (see Herbert, 67).

time. It is only in and as the *encounter* between the cascading photons from the radiating Moon and the receptor sites on your retina that the Moon appears, and "is," as we experience it—that is, as round, and in this instance suspended like a large paper lantern just above the darkening hills. Another example, this time not involving two bodies (the Moon and the organism, or a photon and a receptor molecule) but consciousness and chemistry (or mind and organism), comes from the new interdisciplinary field of psychoneuroimmunology: the intentional cultivation of a certain mood—say, joyful confidence— or of an idea (essential goodness) or image (white light) generates a corresponding neurochemical profile that is in turn associated with improved functioning of the immune system. The general principle at work here is actually the same as that in the now well-studied, though no less mysterious, relaxation response, where blood pressure, heart rate, and brain-wave activity are subject to alteration through the action of conscious intention (there is also, of course, the equally mysterious placebo effect).

A more complete accounting of our cosmic solidarity would involve a consideration of the nature of life. Before leaving the realm of physics, however, I would draw particular attention to the evidence in favor of what is called *the anthropic cosmological principle*, which holds that life in general and human existence in particular are somehow integral to the deep structure or organization of the cosmos. The evidence centers around the exceptionally fine-tuned values of certain physical constants. As James Gardner explains:

> every aspect of the evolution of the universe—from the birth of galaxies to the origin of life on Earth—is sensitively dependent on the precise values of seemingly arbitrary constants of nature like the strength of gravity, the number of extended spatial dimensions in our universe, and the initial expansion speed of the cosmos following the Big Bang. If any of these physical constants had been even slightly different, life as we know it would have been impossible.[169]

As for the nature of life, both in itself and as an expression of cosmic solidarity, we might appeal to the view that, to quote the title of a book by Stuart Kaufman, life is "at home in the universe," which is to say that it is of the very nature of nature or the cosmos to produce life, and therefore ourselves as living beings. As I briefly alluded to in the preceding chapter in connection with Gaia theory, life has created and maintains the planetary atmosphere as we know it and depend upon it. And we could go further: following the Romantic philosophers of nature and such new paradigm figures as Bateson, Varela, and Morin in our own time, we could say that it is the complex nature of life that holds the key for understanding the life of nature as a whole. For it is with life—with such phenomena as morphogenesis, metabolism, healing/repair, reproduction, and speciation—that the cosmos reveals itself most explicitly not only as dynamic and evolutive, but also as sentient, responsive, and purposeful. This is most obviously the case with human life, in which evolution or the cosmos becomes conscious of itself as such. More locally and critically, however, it is also with human life that, for the first time in sixty-five million years, the planet is now faced with a general biospheric crisis.

2) Human Unity

Our solidarity with the cosmos already implies or grounds the ideal of human unity at the most fundamental level. Whatever might seem to separate us from one another, we now know that we are among the youngest siblings of a cosmic family whose ancestry stretches back beyond the first *Homo sapiens sapiens* (some 200,000 years ago), through the origins of life on planet Earth (perhaps as early as 3.8 billion years ago), to the primal "flaring forth" and, in the beginning, the mysterious "Singularity" (fifteen billion years ago). The surprising proximity of humans to our nearest cousins on the tree of life was recently revealed with the initial findings of the human genome project, which has shown that we share some ninety-seven percent of our

genes with the genomes of the rest of the hominid family (and, perhaps even more surprisingly, we share fifty percent of our genes with
the lowly worm, *Caenorhabditis elegans*). At the same time, focusing
in on the mere (but obviously significant) three percent that we do not
share with the other primates, current research in the science of mitochondrial genetics suggests that the billions of humans now populat-
ing the planet are the descendents of a single woman—known as the
"mitochondrial Eve"—who lived approximately 150,000 ago in Africa.*
Remarkably, it would appear that, some 80,000 years later (about 70,000
years ago), most likely due to a sudden ice age brought on by the eruption of a super volcano in Indonesia, the total human population was
reduced by half to no more than 10,000 individuals. Everyone alive
today would therefore be the descendent of one of these early survi-
vors of global catastrophe.

Assuming an African genesis, the initial human diaspora began
about 130,000 years ago, covering Africa and Eurasia and crossing the
Bering Strait over the next 30,000 years, reaching Australia and New
Guinea about 40,000 ago, and finally Polynesia just a few thousand
years before the present era. This diaspora facilitated the expression of
a wide range of physical, and especially linguistic, differences among
humans—the standard estimate is that there currently exist over 6,000
different languages. (These can perhaps be grouped into around two
dozen language families.) It is impossible, of course, to know how
many languages were lost during the many thousands of years preceding the historical period. As for differences of "race," contrary to common belief (and long-standing prejudice), it has been demonstrated
through genetic research that "most physical variation, about ninety-
four percent, lies *within* so-called racial groups.... This means that
there is greater variation within 'racial' groups than between them." [170]

---

* This "Eve" probably had about 20,000 genetically similar contemporaries,
none of whose descendents, however (or too few to be detected), have
survived to the present.

Whatever the physical differences, whether within or between such groups, there remains a fundamental genetic sameness that allows for the biological pairing between any fertile man and woman on the planet. This genetic sameness can be taken as an indication of a more general anthropological unity that includes the full range of human expression. In spite of the diaspora, as Morin says,

> and in spite of physical differentiations in cultures, and the fact that languages became mutually unintelligible, rituals and customs incomprehensible, and peculiar beliefs unyielding to one another, everywhere there has been myth; rationality, strategy, and invention; dance, rhythm, and music. Everywhere there has been— unevenly expressed or repressed according to cultures—pleasure, love, friendship, anger, and hatred. Everywhere there has been proliferating imagination and, however different their recipes and proportions, always and everywhere there has been an inseparable mixture of clear-headedness and folly.[171]

Affirmation of human unity through the recognition of such anthropological constants does not, of course, entail the denial or minimizing of cultural and individual differences. As in the case of our cosmic solidarity, the unity in question is *complex,* which means that it both supports and is supported by virtually endless potentials for variation, modulation, or inflection. Without these potentials, our species would not have been able to adapt and respond creatively to the differences in habitat and overall circumstance that it encountered throughout its planetary wanderings. As the work of Jung, Mircea Eliade, and Campbell has done so much to reveal, these potentials have as much to do with the creative depths of the human soul or psyche, with the specifically human quest for meaning, as they do with any strictly genetic or biological plasticity. While all three of these figures argued for a common archetypal foundation to human nature, it was precisely through sympathetic engagement with the world's diverse cultural expressions that they were able to intuit the essential unity behind or within the multiplicity—"the hero with a thousand faces," as Campbell put it. It was Jung, however, who most clearly saw the complex holistic character of the archetypal landscape and, as we have seen, pointed the way to a psychology for the Planetary Era with his concept of the Self as *complexio oppositorum*. The wisdom of this insight into the nature of the Self, though continuous with the various strands of mystical and esoteric speculation running from the great sages of the first Axial Period all the way to Hegel and beyond, could not be more relevant to our current moment. With the persisting tendency for nations and ethnicities in conflict to regard each other, as Jung said decades ago now, as "the very devil,"[172]* there is a need as never before for an understanding of human identity as constituted in and by its relation to difference, for a mirroring of self in the other, and of the other in oneself.† Without significant progress in the cultivation and spread of such an understanding within the next decade or so, there are only the dimmest of prospects for the future of the Planetary Era.

3) Radical Interdependence

Human beings have always recognized dependence as a fundamental fact of existence—dependence on plants and animals for physical survival, on our mothers or caregivers in infancy, on the weather, on the beneficence or the whim of social superiors, on the will of the gods or of God, on sheer luck, and of course on the good will of friends and strangers alike. And though there have doubtless always been tyrants and cruel, emotionally stunted or otherwise insensitive people incapable of seeing the mutual inclusion of self and

---

* Echoing Reagan's description of the U.S.S.R. as "the evil empire," we have George W. Bush's more recent references to certain countries belonging to "the axis of evil."

† This Self is what Hegel also calls Spirit (*Geist*), which he at one point defines as "the I that is We and the We that is I" (see Kelly, 1993 a, 85).

other—and therefore incapable of realized compassion—it was not until the Planetary Era that this human failing took on truly global proportions. Three related factors can be singled out in this connection, all of which have played determinative roles in the emergence of the Planetary Era.

To begin with, there is the rise of modern science and technology, which brought with them unparalleled means of control over nature (and everything else considered, or made into, an "object"). Second, and partially as a result of technological advances (in food production, in navigation and the production of arms, and especially following the Industrial Revolution), the world saw the transformation of nation–states into global empires and eventually superpowers. Finally, there is the phenomenon of global capitalism, which, though associated with the concentration of wealth (or the means of production) in the hands of the few (whether counted as individuals or nation–states), is explicitly transnational—that is, imperialistic—in attitude and aspiration.* The worldviews of modern science, of the modern nation–state, of global capitalism, and the dominant paradigm in which they all participate are organized in large part around the two related impulses toward pathological autonomy and strategic dominance, both of which involve a corresponding denial of dependence. In each case, moreover, the other is not viewed as a true self, but as mere object or "data," as commodity, as a potential enemy or "alien," or as "primitive" or "undeveloped."

The viability of the dominant paradigm has been challenged, however, and its catastrophic potentials exposed, with the emergence and intensification of a series of planetary crises: it is by no means certain that we have seen an end to world wars—invasions and occupations

---

* While the capital-driven industrial revolution may have raised many from the general population above the level of mere subsistence, it also created a new class of wage slaves. Another consideration here is that pre-industrial subsistence economies were at least sustainable, whereas those of industrial growth society are not.

continue alongside the new wave of global terror; the world economy (with its reliance on new markets and dwindling nonrenewable resources) is increasingly precarious and uncertain:* the pressures of population and runaway consumption threaten the integrity of the biosphere (deforestation, pollution, global warming, the extinction of species). The modern ideals of autonomy (atomistic individualism, national sovereignty, corporate entitlement) and techno-industrial progress (so-called development, limitless growth), though instrumental in ushering in the Planetary Era, now threaten its further actualization.

It is these same crises, however, which are now forcing us to acknowledge the fact of our radical interdependence. To begin with, between humanity and the rest of the biosphere: one might think that our dependence on the health and bounty of the living Earth goes without saying. In fact, however, it is only in modern times that this dependence began to be less purposely and regularly affirmed in ritual, prayer, the telling of myths or sacred stories, and the common actions and gestures of daily life. (The saying of grace at mealtimes is one vestige of such an affirmation.) The prodigious successes of the Industrial and (in retrospect, ironically misnamed) "green" revolutions (ironic because of the devastating effect of chemical fertilizers, pesticides, and herbicides), the seemingly endless series of technological innovations, and of course the triumphalist ideology of limitless progress, were able for a while to create a false sense of all-sufficiency. The near disappearance from over-fishing of the once abundant north Atlantic cod, the drastic diminishment of wild salmon, and the threat to the maple-sugar industry because of acid rain were some of the first signs that something was amiss. At the time of writing, U.N. scientists warn that one in six countries face food shortages and many the threat of irreversible desertification.[173] There is a global

---

* Since I first wrote this passage in 2006, just how precarious became glaringly obvious with the global financial crisis precipitated by the collapse of unregulated credit markets.

consensus on the dangers of climate change and on the fact that we have already passed the peak of world oil production (the projections are that, at current rates of consumption, there are only a few decades left of usable oil—and it is upon oil, of course, that so much of our current mode of life depends). To get an overall sense of where we stand, we need only reflect on the findings of the 2008 Living Planet Report,[174] which estimates that our global "ecological footprint" is already at least twenty-five percent greater than the Earth's productive capacity. Even assuming the ability to maintain current rates of production (which is a big assumption, given the possible effects of global climate change, peak oil, and other unforeseeable disruptive factors), our limited and straining Earth clearly cannot sustain current levels of consumption.

The denial or disregard of our dependence on the rest of the biosphere has gone hand in hand with a systematic blindness to the devastating effect we have already had on the planet. The same Living Planet report "shows a forty percent decline in terrestrial, freshwater, and marine species during 1970 to 2000." The consensus among the world's leading biologists and ecologists is that, if present trends continue, more than "one half of all species of life on Earth will be extinct within one hundred years."[175]

The radical interdependence of humans and the rest of the biosphere is mirrored by the radical interdependence among all humans on the planet. In the first place, this has to do with the fact that human actions take place within, or interact with, essentially continuous and planetary media—notably, the atmosphere and the hydrosphere—which leads to the virtual impossibility of containing harmful substances released into the environment. Here the most serious threats come from nuclear contamination (whether from the explosion of nuclear bombs, failed reactors, accidents involving stored nuclear waste, or the use of depleted uranium) and chemical pollution (from greenhouse gases to PCBs and acid rain).

Our ecologically mediated interdependence is mirrored at the
level of geopolitics. This was first made catastrophically evident in the
last century with the World Wars, which grew out of, as they restruc-
tured, the complex interdependencies among the planet's nation states
and broader alliances. The struggle between the two largest nuclear
superpowers has given way to the current configuration, which is
simultaneously more unipolar and more polycentric. (With the still-
dominant, though increasingly challenged, United States in various degrees of alliance with the E.U., Canada, Japan, Israel, and the
Indian subcontinent, on the one hand, and in strained to antagonistic relations with Muslim–Arab countries, Russia, China, and parts
of Central and South America, on the other.) From the dominant
realpolitik point of view, our interdependence—to the extent that it
is recognized at all—is based on the need for security (for instance,
the fight against global terror) and economic self-interest (so-called
free trade agreements)—in other words, on fear and greed. There
is, however, a growing movement that chooses instead to recognize
our interdependence as an expression of our collective higher-order
needs (in the sense of Maslow's hierarchy of needs, which proceed
from basic physiological and security needs, through belongingness
and self-esteem to self-actualization and self-transcendence). Perhaps
the most significant expression of this movement is the *Earth Charter*,
which begins with these words:

> We stand at a critical moment in Earth's history, a time when
> humanity must choose its future. As the world becomes increas-
> ingly interdependent and fragile, the future at once holds great
> peril and great promise. To move forward we must recognize that
> in the midst of a magnificent diversity of cultures and life forms
> we are one human family and one Earth community with a com-
> mon destiny. We must join together to bring forth a sustainable
> global society founded on respect for nature, universal human
> rights, economic justice, and a culture of peace. Toward this end,

it is imperative that we, the peoples of Earth, declare our responsibility to one another, to the greater community of life, and to future generations.[176]

4) Spiritual Liberation

If the 1960s Counterculture ushered in a new age of social, psychological, and artistic liberation, the Planetary Era in the making, if it is to fulfill its deepest potential, will involve what might be called a *spiritual* liberation—and this in two senses. First, a liberation of religion or spirituality from the cancer of fundamentalism, and more generally from the deeply entrenched obstacles to greater mutual understanding and creative dialogue among the world's religious/spiritual traditions. The second, more positive sense of spiritual liberation concerns the liberatory potential of religious/spiritual traditions, symbols, and experiences. To begin with the first sense: the problem of fundamentalism obviously cannot be solved through an appeal to higher values, reason, or common good will. Fundamentalist mindsets and their associated worldviews are notoriously impervious to challenges, however cogently articulated and however forcefully or gently administered. And we can readily understand why, once we realize the degree to which these mindsets are sustained as much from within as from without.

Freud was the first to uncover the collusion in depth between the religious (and here I would read *fundamentalist*) imagination and unconscious psychological complexes rooted in childhood experience. To the extent that the psychoanalytic insight holds, we can expect no serious progress in the anti-fundamentalist agenda without prior successes in the domain of psychotherapeutics, which, as we now know, must extend beyond the consultation room to the practices of early childhood education, child rearing, and even child birth. At the same time (as we saw in chapter 9), the social dynamics of fundamentalism are bound up with the forces of global capitalism, which, in its current American configuration, has been sustained by the alliance between Big Oil and the Religious Right (and, correlatively, by the crusade against the so-called axis of evil). The problem of fundamentalism, in other words, is highly complex. In this case we see how the question of mindset or worldview is holographically and recursively embedded within the structure and dynamics of the individual psyche, of the family, of civil society, and of international relations. The complexity of the relations among all of these terms lies in the fact that each of them contains and generates the others, while simultaneously being contained and generated by them. Oppressive social relations, for instance, reinforce unhealthy family dynamics, which individuals then internalize as unconscious personality traits. These traits, and such associated defenses as projection and splitting, serve to perpetuate oppressive social relations.

As for the second, more positive sense of spiritual liberation involving the transformative potential of religious/spiritual traditions, symbols, and experiences, I have already drawn attention to the connections between the *holy* and the cognate terms, *healing* and *wholeness*. According to Jung, it is the special virtue of "living" religion, or more generally the "symbolic life," to put individual and collective consciousness in touch with the life-sustaining, meaning-full depths of the archetypal psyche. Clearly, however, some archetypal potentials are more wholesome than others, as Jung himself recognized in his analysis of the widespread possession of the German psyche by the god-figure Wotan leading up to and during World War II. In our own times, one need only contemplate the horror of the suicide bomber. However one might want to deny any connection between the killing or maiming of innocent people and a genuine manifestation of the religious or spiritual impulse, the case can be made that such actions are not possible without some particularly potent archetypal undergirding.

How are we to know whether or not a given religious or spiritual phenomenon speaks with the voice of the Self? It is not enough to point to a mere structural or phenomenological suggestion of wholeness. The Nazi swastika, for instance, though unambiguously both numinous—and therefore "holy"—and mandalic (quaternity and sun wheel)—and therefore suggestive of wholeness—was and remains bound up with a titanic lust for power and a genocidal fury that militate in the strongest terms against the kind of healing and wholeness that Jung wishes to associate with the deepest telos of the Self. While there can be no certainty in such matters—a point I shall return to in a moment—we can be assisted in our deliberations if we recall that the Self, though considered the central archetype of the collective unconscious, is more properly conceived as the psyche in its totality, including both consciousness and the unconscious. This means that numinous encounters with the archetypal psyche, if they are to serve the actualization of wholeness, must be yoked to the discriminating ego and its potential for reflexive awareness. It is only within the orbit of such awareness that one can hope for any possible progress, however halting and provisional, through the complex and dangerously charged terrain of human religious/spiritual expression.

Obviously, potential conflicts among the world's many spiritual traditions are more severe to the extent that, as in religious fundamentalism, a given tradition maintains an exclusivist or chauvinistic stance relative to other traditions. Even in the absence of such a stance, however, the sheer richness and diversity of these traditions would seem to resist the kind of mutual understanding necessary for the fruitful unfolding of the Planetary Era. Among the many attempts to articulate perspectives that might serve such an understanding, I would point to two dialogically related approaches, both of which have been pursued in the context of transpersonal theory. The first could be described as *essentialist* or *perennialist* and is perhaps best represented in the work of Wilber, much of which, as we saw in the previous chapter, is informed by the idea of the perennial philosophy (and psychology). The main claim of this approach is that there exists a fundamental, unitary structure beneath the surface multiplicity of religious/spiritual forms. A common metaphor here would be that of many paths leading to the summit of the one mountain. At the same time, however, most perennialists—and this is certainly the case with Wilber—also maintain that the deep structure is organized hierarchically (the "Great Chain" or, in Wilber's eventual modification, the Great "Nest" of Being), which means that some forms are closer to the summit than others. For instance, in Wilber's estimation, the mystical monism of Vedanta, or the non-dualism of Tibetan Mahamudhra or Dzogchen, is considered more realized than, and therefore superior to, non-mystical theism, indigenous animism or panpsychism, or to various kinds of nature mysticism.

Wilber's essentialist approach has been roundly criticized by Jorge Ferrer, whose revisioning of transpersonal theory takes full advantage of the postmodern turn in philosophy and religious studies. The tendency here is to problematize the truth claims of totalizing (and potentially totalitarian) metanarratives and to stress the creative, constructed, context specific, and therefore inherently pluralistic and generally heterarchic (that is to say, more horizontally than vertically organized) nature of spiritual expression. In place of the mountain with many paths, Ferrer proposes the image of the ocean with many shores. Summarizing his proposal, he writes:

> What the mystical evidence suggests is that there are a variety of possible spiritual insights and ultimates (Tao, Brahman, *sunyata*, God, *kaivalyam*, etc.) whose transconceptual qualities, although sometimes overlapping, are irreducible and often incompatible (personal versus impersonal, impermanent versus eternal, dual versus nondual, etc.).... Although these spiritual ultimates may apparently share some qualities (e.g., nonduality in *sunyata* and *Brahmajñana*), they constitute independent religious aims whose conflation may prove to be a serious mistake.[177]

While both Wilber and Ferrer see their work as offering a middle way between the conflicting positions of exclusionary fundamentalism, on the one hand, and anarchic relativism, on the other, they nevertheless each tend to favor one side of, even as they strive to mediate, a series of opposed terms: hierarchy (or holarchy) versus heterarchy; absolutism versus relativism; monism versus pluralism; and as we have seen, essentialism versus constructivism.[178] All of these oppositions can be seen as manifestations of the more fundamental relation between identity and difference, the complex nature of which, as we have seen repeatedly, has played such a generative role in the birth and transformation of the Planetary Era. As with the other planetary ideals, the actualization of spiritual liberation will draw upon, as it cultivates, a planetary wisdom that embodies the principle of complex holism (by whatever other name).

An insightful vision of this kind of planetary wisdom is Ewert Cousins's notion of the "Second Axial Period." As with the first Axial Period, the second is happening simultaneously around the Earth. In contrast to the first, however—whose focus was on the transcendent potentials of individual consciousness—the Second Axial Period is explicitly centered on the cultivation of planetary consciousness. "In this Second Axial Period," writes Cousins,

> we must rediscover the dimensions of consciousness of the spirituality of the primal peoples of the pre-Axial Period...this consciousness was collective and cosmic, rooted in the earth and the life cycles.... Having developed self-reflective, analytic, critical consciousness in the First Axial Period, we must now, while retaining these values, reappropriate and integrate into that consciousness the collective and cosmic dimensions of the pre-Axial consciousness.... This means that the consciousness of the twenty-first century will be global from two perspectives: (1) from a horizontal perspective, cultures and religions are meeting each other on the surface of the globe, entering into creative

encounters that will produce a complexified collective consciousness; (2) from a vertical perspective, they must plunge their roots
deep into the earth in order to provide a stable and secure base
for future development.[179]

Notice how the kind of spiritual liberation Cousins sees in the Second
Axial Age not only involves the recognition of human unity (his horizontal perspective) and cosmic solidarity (his vertical perspective),
but also the triphasic structure that marks any truly complex holistic
process (in this case, the Second Axial wisdom as the creative synthesis of the "primal" [identity] with the "critical" [difference]).

Cousins' vision of the Second Axial Period raises some fascinating questions, which I will simply pose here: Does not the very invo-
cation of a "post" Christian era presuppose the millennial schema of
the Christian worldview—*a New Testament by another name*? Or is
the Christian symbolic matrix, despite its generative function as the
Great Code for the emergence of the Planetary Era, destined to be
fully overcome, like the shift from the pupa to the imago in the life
cycle of the butterfly? If Christianity served as the mother butterfly,
with the rise of the West and the emergence of the Planetary Era as
the pupa or chrysalis stage, can we expect the emergence of a new
butterfly—the (re)birth of a new, and for the first time truly planetary, World Soul? For my part, I would second Thompson's affirmation that a new Planetary (Wisdom) Culture would not

> mean a national or ethnic culture that becomes triumphant, but a
> new emergent domain that is bounded and energized by the mem-
> brane of the Earth's biosphere and is characterized by new expres-
> sions of art and science [and religion?] that are not restricted to
> historically defined ideologies or groups. The ethnicities within
> become like the organelles within a cell. Like mitochondria, they
> can keep some of their ancient DNA and their intracellular unity,
> but their expressions now serve to energize the metabolic processes
> of the larger planetary cell.[180]

We can at least say this much: Our aspiration toward a planetary wisdom must involve the renunciation of certainty—or at least of that kind of certainty that cannot countenance contradiction and knows truth only by exclusion. "Contemporary reflection," writes Morin, "must begin with the consciousness of the limits of knowledge, not so as to enclose itself within these limits, but in order to become a sentinel of the unknown and a satellite of the inconceivable."[181] It is a question, as Cusa saw so well half-a-millennium ago now, of a *learnèd ignorance*, of including the unknown within our knowing. This wise ignorance is not only learned—informed as it is by the many and complex insights we have had occasion to consider throughout these pages—but must be *learned*. It is a coming to know that we don't know, yet also a growing knowledge-as-intimacy with the unknown, an encountering of what ever overflows the grasp of the certainty-seeking ego. Such a wisdom, while it relativizes the ego, does not simply negate or obliterate it! —since it is only from the perspective of the finite ego that knowledge of ignorance, or consciousness of the unconscious, can arise in the first place. The vision of such a wisdom is keenest in the chiaroscuro between night and day.

Much of the wisdom—if wisdom there be—that has inspired this book has come from tracking the flight of the Owl of Minerva, which, as Hegel famously put it, "takes flight only when the shades of night are gathering." This kind of wisdom

> does not appear until reality has completed its formative process, and made itself ready.... When philosophy paints its grey in grey, one form of life has become old, and by means of grey it cannot be rejuvenated, but only known.[182]

While we should certainly continue to draw insights from this noble bird and its learned trackers, in these dark times there is perhaps another kind of wisdom to be gleaned from the common sparrow, the most widely distributed wild bird on the planet. Now threatened, along with all wild birds and the rest of our Earth community, by the worsening global ecological crisis, there is something even more precious and poignant in its simple song, which, for now at least, it offers up each morning to the first light.

EPILOGUE

WHAT would it mean for us truly to come home? How would we know we had arrived? For the indigenous mind, home was a definite place (often as small as a village, sometimes as large as a bioregion), the place of one's ancestors, the stories of whose founding actions served as guidelines and regulators for sustainable living, as stores of practical knowledge and wisdom for both individuals and the community. This place was imagined as the center of the world, just as one's clan or tribe provided the type for what it meant to be human. Despite the archetypal themes underlying them, these myths—and thus also the typical human that they defined—were as numerous as the groups within which they originated. Still, "home"—or rather, the many homes—were rooted in the life of the Earth, in the cycles of the seasons and the constant rhythm of the passing generations.

With the birth of the historical period, and especially following the first Axial Period, one sees the emergence of new, more universalizing myths and cosmologies. A new tension is also introduced: on the one hand, a complexification of the older, Earth-based orientation, now harmonized within the life cycles of cultivated plants; on the other hand, a preoccupation with notions of the afterlife, with a transcendent beyond most commonly associated with the heavens—whether with the Sun or the stars, in either case suggesting a realm of eternity or permanence in contrast to the changeable and death-ridden human condition and of all of life below the sphere of the Moon.

As we have seen, among the many mytho-cosmological visions that arose on the planet following the first Axial Period, it was the biblical mythos that happened to serve as the Great Code for the eventual birth of the Planetary Era. For a thousand years prior to this birth, the West envisioned its true home along the lines of the New Jerusalem of Revelation or the beatific vision of Dante's *Paradiso*. By the sixteenth century, however, Renaissance humanism, the Copernican revolution, the Protestant Reformation, and the "discovery" of the New World all point to that epochal shift in axes from the vertical to the horizontal, from Heaven to Earth, which characterizes the transition to the Planetary Era.* Idealized in visionary utopias— Bacon's *New Atlantis,* Andrae's *Christianopolis,* Campanella's *City of the Sun,* More's *Utopia*—the New Jerusalem is now to be sought this side of the Millennium, and notably on the other side of the Atlantic.

This impetus culminates in the three great early revolutions (British, American, and French) that led to the first modern democratic republics, two of which having their own versions of a sacred-secular trinity (life, liberty, and happiness for the Americans; liberty, equality, and fraternity for the French). In little more than a century, however, these and other fledgling states are swept up in the planetary paroxysms of two world wars, the second of which, with the onset of the nuclear age, unites the planet beneath the specter of human-generated apocalypse. A mere quarter-of-a century later, the world has awakened to the growing ecological crisis, which, in the first years of this new millennium, holds the real possibility of a mass extinction of species and generalized ecological and civilizational collapse.

---

* While the Copernican revolution displaced the Earth in favor of the Sun as the center of the observable cosmos, as we saw in chapter 4, this displacement was the first and necessary step toward the articulation (realized by Newton) of a universal physics (unifying Heaven and Earth and allowing for an unprecedented manipulation of the material world). The pre-Copernican Earth, though spatially central, stood at the lowest level of the ancient and medieval vertical hierarchy. In the post-Copernican cosmos, this hierarchy is leveled and all matter, celestial and terrestrial, now lies on the same (ontologically) horizontal plane.

At the beginning of the sixth century of the Planetary Era, we are faced with the most extreme and fateful of contradictions. Alongside the threat to the biosphere, and stimulated by this threat, and again both despite and in compensation for the newly aggravated danger of nuclear war and deepening ideological and socioeconomic divisions, there are such manifestations of a truly planetary, because potentially sustainable, Earth Community as the Earth Charter, increasing international cooperation (however hobbled), and the growing populist and progressive influence of internet-facilitated democracy. There are also the hundreds of thousands of (as yet non- or barely associated) progressive organizations around the planet that Paul Hawken has identified as the largest social movement in history, and that I call the movement for Global Solidarity. When the New Paradigm and New Age (broadly defined) groups that focus more on the psycho-spiritual dimensions of individual and global transformation are added into the equation—some notable examples include the *Institute of Noetic Sciences* (IONS), the *International Transpersonal Association,* the *Scientific and Medical Network*—the movement, already the largest ever, suddenly doubles in size.

As is the case with the swelling number of groups identified by Hawken, there is as yet no widely used name for this largest ever social movement.* For our purposes, we can identify it as the movement toward a *planetary wisdom culture.* This is not to say that all or even most of these groups would embrace or enact the principle of complex holism articulated in the last chapter. The majority of them would, however, affirm some version of the four planetary ideals of cosmic solidarity, human unity, radical interdependence, and spiritual liberation with which I illustrated the nature and virtue of this principle. This emergent planetary wisdom culture is the latest flowering— and really the first to surface on a planetary scale—of the periodic

---

* Paul Ray and Sherry Anderson have suggested the term *cultural creatives* to describe this portion of the population (see Anderson and Ray, 2001).

countercultural impulse that, as we have seen, has played such a key role in the birth and transformation of the Planetary Era. Hawken invokes Gary Snyder's image of "the great underground" to characterize this impulse. Its lineage "can be traced back to healers, priestesses, philosophers, monks, rabbis, poets, and artists 'who speak for the planet, for other species, for interdependence, a life that courses under and through and around empires."[183] This time, however, the movement cannot go under—unless, that is, we are prepared to see civilization itself and the biosphere as it has existed for many thousands, if not millions of years, go under along with it.

If a truly planetary wisdom culture does succeed in fully emerging and stabilizing itself, there will be much that *must* go under: business as usual, industrial growth society, unchecked corporate rule, unsustainable modes of production and consumption, the roguery of nations, to name the most obvious. We could also say that what must go under is *Empire*—a term, following David Korten's usage, that refers to "the hierarchical ordering of human relationships based on the principle of domination." More particularly, the spirit of Empire "embraces material excess for the ruling classes, honors the dominator power of death and violence, denies the feminine principle, and suppresses the realization of the potentials of human maturity."[184] The antidote to Empire, according to Korten (who adopts the phrase from Brian Swimme and Thomas Berry), is *Earth Community,* which stands for "the egalitarian democratic ordering of relationships based on the principle of partnership." The spirit of Earth Community "embraces material sufficiency for everyone, honors the generative power of life and love, seeks a balance of feminine and masculine principles, and nurtures a realization of the mature potential of our human nature."[185]

To describe the shift toward Earth Community, Korten follows Joanna Macy's evocation of "The Great Turning," an image that harmonizes with the spiraling character of the fundamental pattern over the last two thousand years, as well as with the promise of a Second Axial Period. If we succeed in making this Great (that is, planetary) Turning, "the great underground" will no longer need, or be forced, to go under. In this sense, the countercultural impulse—or the core values it has embodied—will cease being countercultural, since they will no longer be defined in opposition to the dominant culture. The Great Turning thus coincides with the final turn of the spiral, at once the *kairos* or unique and opportune moment, and the *eschaton* or terminus of the multiple converging arcs leading to and through the Planetary Era. As we saw in the last chapter, we can expect the presence of this extended moment to be intensifying over the next decade or so. Futurists point to the imminence of a technological Singularity—Ray Kurzweil, most famously, has fixed its arrival to around 2035—which promises sweeping revolutions not only in communications, computation, energy use, bioengineering, and applied consciousness research, but in the ability to manage more effectively the physical complexities and challenges of the Planetary Era (ecosystem and resource management).[186] The Singularity theorists, however, generally neglect to factor in the economic and political realities, let alone the deeper paradigmatic assumptions of Empire that drive these realities. In any case, we may not have until 2035. At least with respect to global climate change and the current mass extinction, we would seem to have about a five-year window (from the time of this writing), centering therefore around the widely charged 2012 (the *eschaton* of the Maya calendar's "Great Count" of roughly five thousand years), before the tipping points are irrevocably passed.

There is neither stopping nor turning back. Though it is impossible to say with certainty how those on the other side of the Great Turning—the sustainers of a planetary wisdom culture—will name or come to define themselves, we can expect their founders and the transformative communities they inspire to share certain essential elements or qualities with all preceding countercultural turns of the tightening spiral: something resonant with the charismatic communalism of the early Christian community, with the liberatory aspirations of the Joachimites and spiritual Franciscans, the theocosmic vision of the Renaissance magi and utopians, the organicist and re-enchanted worldview of the Romantics and Idealists (and the esoteric streams from which they drew), and the countercultural breakthroughs of the last century, especially those associated with the sixties counterculture and the New Paradigm movement that followed. There will also, no doubt, be novel manifestations of the emergent planetary spirit that, though they can be imagined and even cultivated, will only be definitively recognized after the fact. "Beneath the crust of visible reality," writes Morin, "there is a subterranean and occult reality that will emerge later but that remains completely invisible to the realist." [187]

Still, from the simultaneously open and bounded perspective of that learned ignorance by which the new spirit must be guided, we must face the possibility that the Planetary Era "may...come to naught before it has even begun to bloom. Perhaps humankind's struggles may lead only to death and ruin." Morin adds, "However, the worst is not yet certain, and the game is not yet over. In the absence of any certainty or even probability, there is the possibility of a better world." [188]

Appendix I

Two Summary Diagrams

THE first diagram, titled "The Principal Arc with Spiral" (following page), represents the tightening and accelerating spiral within the larger arc that begins with the First Axial Period and ends with the Second. The rise of the West, the birth of the modern and with it of the planetary, are seen as rooted in the Christian tradition, itself a creative hybridization of earlier Axial traditions. The modern mechanistic worldview, which establishes its dominance from the time of the Enlightenment, is periodically challenged by focused surges of countercultural innovation. Despite the association of the New Right and neoconservatism with (generally fundamentalist) forms of Christianity, these expressions of dominant culture retain essential traits of the modern mechanistic worldview: notably, dualism and a drive for power. Each new countercultural surge is not only resonant with the preceding, but as an expression of increasing planetization draws from earlier Axial and pre-Axial or indigenous traditions.

In the second diagram, titled "The Principal Arc as Middle Term between Origin and Goal" (page 163) we see the previous arc, dominated by the rise of the Solar principle, as the middle term or threshold between the Lunar arc: from (the unknown) Origin to the First Axial Period, and the arc of the marriage of the Solar and Lunar: from the Second Axial Period to (equally unknown) Goal. As per Korotayev, prior to the first Axial Period, objective factors (geography, ecology)

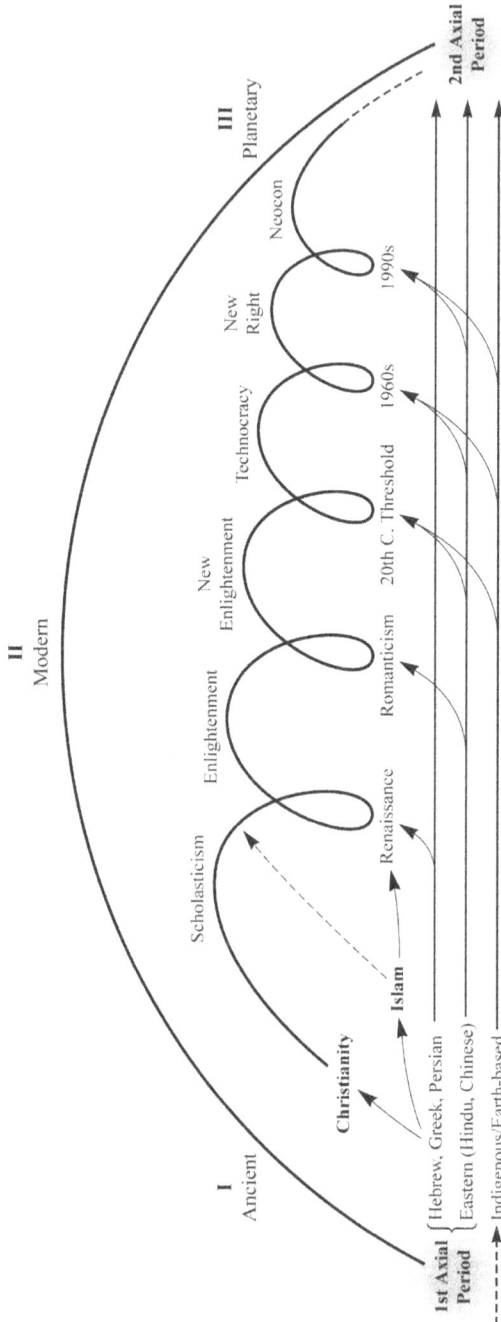

1. The Principal Arc with Spiral

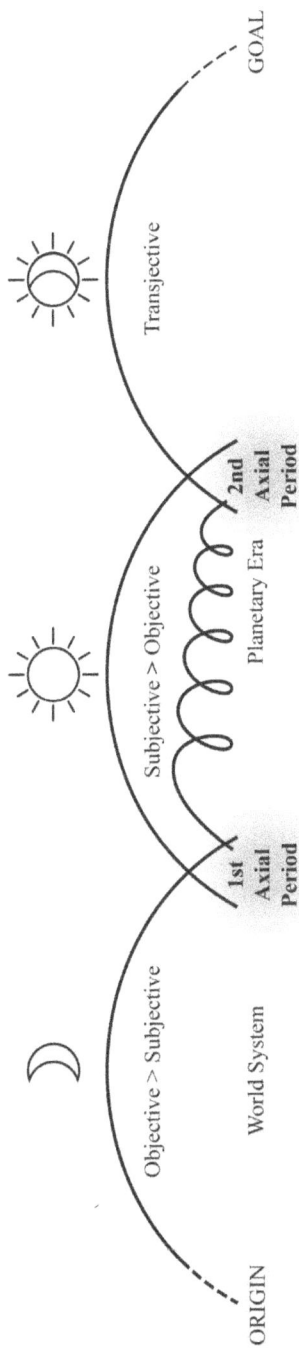

ORIGIN

Objective > Subjective

World System

1st Axial Period

Subjective > Objective

Planetary Era

2nd Axial Period

Transjective

GOAL

2. The Principal Arc as Middle Term between Origin and Goal

predominate over subjective (culture), whereas the reverse is true during the Solar arc (where in our own day, most notably, the fate of the biosphere itself will largely be determined by human choices). In a manner already anticipated by the countercultural inflections of the previous cycles, the Second Axial Period points to a transcendence and transformation of the relation between the objective and the subjective, to a *trans*jective state (for lack of a better word) that mirrors the sacred marriage of the Lunar and Solar.

APPENDIX II:

ARCHETYPAL ASTROLOGY
AND THE EVOLUTION OF CONSCIOUSNESS

1. INTRODUCTION

RICHARD TARNAS has now written two extraordinary books, both as beautiful as they are profound, and both of which have inspired me in the writing of the present work. The first, *The Passion of the Western Mind,* is more than a history of ideas, since it employs as it extends a psycho-philosophy of history, or a theory of the evolution of consciousness, whose perspective follows broadly Hegelian and Jungian lines. The deep structure of this perspective has been described by M. H. Abrams as a "circuitous journey" and corresponds to a particular (and as we have seen, incarnational) inflection of what Joseph Campbell called the *monomyth,* or hero's journey, with its movement out of an original, pre-egoic identity, through an initiatory encounter with death, to a new, more conscious and integrated identity. The second book, *Cosmos and Psyche,* presents a larger body of evidence and a sustained argument for the value and implications of archetypal astrology for an understanding of the complex movements of culture, with a focus on the birth of the modern and the "intimations," as he puts it in the subtitle, of a new worldview.

While the general subject of astrology receives a brief historical treatment in the *Passion,* it is otherwise not explicitly engaged, though in retrospect (that is, after reading the second book) one can see how the archetypal–astrological perspective has informed his presentation of the principal dynamic factors involved in the evolution of the Western mind. *Cosmos and Psyche,* for its part, though it retains the core elements of the Hegelian and Jungian orientations, gives most attention to several planetary cycles that reveal the periodic surges of the corresponding archetypal, cultural, and historical streams, or movements, under consideration. Because his treatment of these cycles is so rich and detailed, it is possible to lose sight of the overall triphasic pattern that *Cosmos and Psyche* shares with *Passion.* This structure is made explicit in the first part of the book, in his discussion of the shift from the Primal (first phase) to the Modern (second phase) worldviews. The new worldview (third phase), embodied and enacted through the archetypal astrological perspective, is not only "intimated," but actually informs the argument throughout. It is because *Cosmos and Psyche* both presupposes and creatively engages the fundamental triphasic pattern that Tarnas is able to wed what might otherwise appear to be two mutually exclusive historical perspectives: the teleological/evolutionary, on the one hand, and the cyclical, on the other. While much of what I will propose is already implicit in Tarnas's written work, or has arisen in dialogue with him, I hope that the following reflections might indicate some fruitful directions for further discussion.

2. *Terra Stella Nobilis*

Tarnas's integration of the astrological perspective with that of the evolution of consciousness involves a recognition of the living Earth as a moving focal point of cosmic meaning, around which the planets (and stars) make their rounds.[189] Despite what might appear to the more literal-minded as a reversion to premodern geocentrism, this idea actually deepens Kant's insight that his proposal for a "transcendental idealism" constituted a second Copernican Revolution, only here, with a third revolution, it is not the human subject that provides the epistemological meta-point of view, but the Earth as a whole (though still mediated, of course, through human knowing). Such recognition allows us to honor to the fullest extent Kepler's insight—*terra stella nobilis*—that the Earth is a noble star, and in one sense the noblest among our solar system, in that it is through human consciousness as it has evolved on Earth that the new structure and unfolding meaning of the cosmos is revealed. Such an honoring would, finally, be more consistent with the simultaneously organicist and anthropocosmic orientations of Hegel and Schelling (and other *Naturphilosophen* influenced by them). For orientations such as these, the Earth (that is, its life processes, and especially as these come to self-consciousness in humans) represents a more concrete, because more complexly organized, and therefore more ensouled, being than do the other planets and stars.* This view, it should be stressed, signals a marked reversal of the dominant view of the ancients, for whom the planets and stars were considered more perfect beings than the Earth (and its four elements).

From the point of view of ancient astrology, though the gods with which they are associated have their myths, the movements of the planets themselves are without a proper history, in the sense of a directed narrative with beginning, middle, and ending. As in modern astronomy as well, their motions are described as the more or less mechanically (and to the ancients, eternally) recurring cycles we see plotted out in the ephemeris. As the ancients conceived of it, only the "sublunary" realm of the Earth and its elements are given over to flux and becoming. The stars and planets were thought to consist of the

---

* A fascinating exception here is G. T. Fechner. Though a student of a student of Lorenz Oken (who was a student of Schelling), and sharing many other elements in common, Fechner proposes that, just as our individual souls effect a conscious merger with the *anima mundi* after death, so the *anima mundi* participates in the larger soul of our Sun, which participates in the soul of the galaxy, and so on. In contrast to Schelling and Hegel, in other words, for Fechner human consciousness is a transitional (rather than the central) middle term in the cosmic evolutionary scheme.

immutable fifth element (the ether) or to be associated with imperishable crystalline spheres. Both ancient and contemporary astrology do indeed see the "influence" of these cycles on, or their correlation with, human affairs. It is my understanding, however—and Tarnas demonstrates this in practice—that these cycles, *by themselves,* cannot account for the overarching trajectory of the history or evolution of consciousness or of culture.

Toward the end of *Passion,* Tarnas characterizes this evolution in terms of a struggle or dialectic between Prometheus (foreshadowing his later treatment of Uranus in *Cosmos and Psyche*) and Saturn and, at a deeper level, between the archetypal masculine and feminine (which, as we have seen, overlaps with the solar and lunar archetypes). In *Cosmos and Psyche,* Tarnas succeeds in giving an account of the evolution of consciousness, especially as this has unfolded over the historical period and into that of the Planetary Era, which attends to the essentially unrepeatable, non-reversible, and non-cyclic (or rather, trans-cyclic, as a spiral) succession of events and *mythemes* that have in fact marked the periods under consideration. One sees in this account that the story of the Earth is not only illumined by, but also illuminates, or *provides the point of view* from which to make sense of the archetypal dance of the planets.

A key consideration here is the discovery of the most studied trans-saturnian planets—Uranus, Neptune, and Pluto—which play such a central role in the archetypal–astrological perspective. Unlike the five visible planets plus the Sun and Moon, the historical origins of whose archetypal meanings is only partially recoverable, the meanings of Uranus, Neptune, and Pluto are known to have been determined initially through extensive study of their positions in natal charts and in mundane astrology (that is, the correlation of their positions with world events). Tarnas, following the pioneering work of Dane Rudhyar, has added to this a kind of synthetic intuition of the overall configuration of the World Soul (*anima mundi*) at the time of their discovery. Uranus, for example, which Tarnas has identified with the mythic Prometheus, reveals itself (among other things) through the breaking into history of the principle of freedom in its distinctively modern form, with its ties to the great political, cultural, and industrial revolutions of the late eighteenth century. The aptness of this identification is confirmed by the fact that one can point to earlier and subsequent moments in the Uranus cycle (in its quadrature alignments with other planets, especially Jupiter, Saturn, Neptune, and Pluto), which also manifest a distinct rising to the surface of the Promethean impulse. The point I want to stress here, however, is that, in this case of the discovery of Uranus—the first of the trans-saturnians, and thus the point of breakthrough out of the premodern planetary cosmic picture where the Earth and human life were in some sense subject to the "influences" of the endlessly repeating cycles of the visible planets—there is something unrepeatable, qualitatively unique, and, one could even say, paradigmatic about the historical moment corresponding to the moment of discovery. It is through the spirit of this unique moment that the meaning of the entire sequence of prior and subsequent uranian moments is revealed.

3. History, Evolution, and the Great Code

In this case, what is revealed is that *freedom has a history*. Again, taking the example of the quadrature alignments of Uranus with Saturn or Neptune or Pluto, without an openness and sensitivity to the kind of teleological–evolutionary perspective proposed by the likes of Hegel, Jung, Teilhard, and Tarnas and in the pages of this book, one might see the rhythmic pulses of the drive for freedom and discovery, but not the entelechy or the longer-range organic patterning of the sequence of these pulses. While it is possible to formulate a philosophy of history or theory of the evolution of consciousness according to a purely cyclic-organismic pattern (as we see with Vico, Spengler, and to a certain extent even with Toynbee), it would seem that an additional perspective or analogy is required to make sense of "the arrow of time," with its associated qualities of radical specificity, irrepeatability, and genuine novelty.

Tarnas's approach to the idea of evolution, of course, is fully informed by the Hegelian–Jungian perspective, which is not only organismic but is also stretched between an Alpha and an Omega (however these are interpreted), with each station in between representing a true emergence of greater depth, complexity, and consciousness. (Though in the more Jungian and Romantic inflection favored by Tarnas and me, the middle, "disenchanted" phase is seen as involving a particular kind of loss as well.) In any case, the appeal to the idea of evolution involves a contextualization of the merely cyclic movements of the planets to the life of the greater organism of the evolving cosmos and of the Earth, both of which have a definite history with a beginning, an unfolding and essentially unforeseeable drama, and presumably some kind of end.

Arthur Lovejoy made famous the understanding of evolution as "the temporalization of the Great Chain of Being." This understanding can be seen as a specific corollary of the more general proposal—as argued by Karl Löwith, for instance—that the modern worldview (which birthed the idea of evolution, and before that the ideal of progress) must be understood as the secularization (and this as an organic metamorphosis) of the formerly dominant Christian worldview. The spirit of this proposal is perhaps best summed up in Blake's lapidary pronouncement that the "Old and New Testaments are the Great Code of Art." In this light, we can see with greater clarity just how the emergence of the modern period and, more important, the Planetary Era, is linked directly to what McNeill termed "the rise of the West," whose guiding mythos, or symbolic matrix, in my reading, has been the biblical and specifically Christian view of time and history. This view, as we have seen, is directed, or at least as irrevocably marked, by a singular Event, the meaning of which can of course be interpreted from a number of perspectives—from the literalist/fundamentalist to the secular/humanist to the more interesting possibilities offered by speculative theology, philosophy, and psychology (as with Teilhard, Hegel, or Jung). Without such explicitly speculative considerations, an astrological perspective like that of the ancients can illuminate and deepen one's understanding of any particular kind of *moment* or *series* of moments of the historical process (with each occurrence of the quadrature aspects in the cycles of the outer planets, for instance), but it cannot, by itself, provide a comprehensive or fully coherent account of the telos of history or the evolution of consciousness. Because it includes such an account, however, Tarnas's revisioning of astrology allows us to see how (in what specific ways, relative to which phase of the fundamental pattern or to which turn of the spiral) the movements of time, evolution, and planetary history have actually played themselves out.

4. From the Anima Mundi to the Weltgeist

A possible critique of *Cosmos and Psyche* concerns the greater attention devoted to Western culture and historical events.[190] If planetary alignments indicate a universal field of archetypal meaning, so the critique would run, one should expect to see this meaning reflected equally in the lives and happenings of people from around the globe at whatever time is being selected. Tarnas actually treats many periods that do seem to satisfy this requirement in a striking way. His treatment of two of these periods is especially compelling. The first has to do with the only triple conjunction of Pluto, Neptune, and Uranus in historical times, from the 580s to the 560s B.C.E. This was the high point of the first Axial Period, which, as we have seen, witnessed the more-or-less simultaneous emergence of many of the world's major religious and philosophical traditions, including the presocratics Thales and Pythagoras and the birth of Western philosophy in Greece, Confucius and Lao Tzu in China, the Buddha and

Mahavira in India, possibly Zoroaster in Persia, and the major Jewish prophets Ezekiel, second Isaiah, and Jeremiah in the Near East. Closer to our own times, there is the Uranus–Pluto conjunction of 1960 to 1972, which coincided with a global eruption of student protests and a generalized mobilization of the youth in major cities around the world, including Berkeley, Chicago, New York City, Mexico City, Paris, Prague, and Beijing. (A similar wave of revolutionary protests swept across Europe in 1848 during the previous conjunction.)

In terms of the main periods considered in this book, these two alignments coincide with the source point of the major arc (the First Axial Period) and the fourth turn of the spiral (1960s). It is interesting to note that all but one of the six turns (ca 1400, ca 1790, ca 1900, ca 1968, ca 2012) coincide with one of the quadrature alignments of the Uranus/Pluto cycle (the fifth turn, ca 1989, coincides with a alignment within the Uranus–Neptune cycle). While it may indeed be the case that, in each of these periods, one can point to social and cultural developments outside of the West that also exemplify the alignments in question, one is arguably justified, when trying to understand the birth and transformation of the Planetary Era, in focusing on the West for the most emblematic and consequential instances of particular archetypal alignments (in this case, those of the Uranus–Pluto cycle) and, I would argue, for instances of particular alignments within the various iterations of an ongoing cycle. As we have seen, the scientific, political, and industrial revolutions associated with the birth and transformation of the Planetary Era all arose in the West. Although there is nothing in the astrological cycles themselves to indicate why this should be the case, the kind of analysis of these cycles provided by Tarnas in *Cosmos and Psyche* does allow for a rich and finely nuanced treatment of the complex turns in the evolutionary spiral of the Planetary Era.

In this light, it could be helpful to distinguish between the movements of what Hegel called the *Weltgeist* (world spirit), which can be thought of as expressing the leading edge or focus of intentionality within the overall trajectory. Once we grant the particular role of the West—Europe to begin with, then especially the United States—in catalyzing the birth of the Planetary Era through the world-transforming combination of the modern secular ideals of progress and freedom with the darker colonial aspirations of the great nation states (some of them destined to become superpowers) and the overwhelming force of the industrial and continuing technological revolutions, one might be justified in looking to the West during the past several hundred years for the most telling events corresponding to any given major planetary alignment. While the World Soul is, as it were, evenly spread across the entire globe, the world spirit—which is the spirit of history itself—would seem to concentrate itself wherever those events are transpiring which will have the greatest consequences for the further unfolding of the overall trajectory. (From a Jungian perspective, one could consider the *Weltgeist* to function as a kind of planetary ego—or perhaps what Jungians refer to as *the ego–Self axis*—while the *anima mundi* would correspond to the collective psyche.*) In this way, not only would one not expect to see equally powerful or significant expressions of a given planetary combination with each of its quadrature alignments, but it could also make sense to focus on the French Revolution and the Romantic–Idealist movement, for instance, which followed in its wake, or on San Francisco during the 1960s counterculture or the fall of the Berlin Wall and the Velvet Revolution of 1989 as the most significant or exemplary manifestations of the particular planetary cycle or alignment in question.

---

* Hegel remarked on the westward movement of the *Weltgeist*. Giving the lie to those who claim that he saw world history ending with the Prussian state, he suggested that the focal point of the *Weltgeist* would likely move out of Europe to the United States, and this more than a century before the latter's rise to the status of superpower.

5. *Freedom and the Hieros Gamos*

According to Hegel, the goal of world history, the deepest passion or longing of the world spirit, is the actualization of the principle of freedom.* We have already seen that a merely cyclical astrological perspective on the Uranus–Pluto alignments, though it might serve to identify a series of historical moments of manifest freedom (and point to resonances between such moments), would not reveal the overarching pattern or trajectory of history or the evolution of consciousness within which the cycle is embedded. Seen within the context of this trajectory, a few critical moments do serve to throw this pattern or trajectory into relief. While the early Greek ideals of autonomy and democracy are clearly relevant to the story, Hegel claims that it is Christianity (with its unique stress on the absolute value of the individual) that introduces the principle of universal freedom into the world-historical process. It is, in effect, by tracking the working out of this principle over the centuries (which is precisely what Tarnas does) that the overall pattern or trajectory is revealed. Following the implantation of the principle in mythic form with the teaching of the gospels, two of the most significant subsequent events are the Protestant Reformation—which absolutizes the freedom of individual conscience (under God)—and the American and French Revolutions, which establish the now secularized ideal of freedom as the universal standard of the modern state. Needless to say, none of these events brought about the full realization of the principle of freedom. The main point, however, is that the movement of history over the last two millennia—and in this case the particular movement of the actualization of freedom—manifests a definite trajectory, marked by a number of critical phase shifts, the specific character and telos of which is invisible

---

* As Richard Tarnas reminded me, Rudolf Steiner would say both freedom and love, which would actually be consistent with the more Romantic phase of Hegel's early thought.

to a perspective that restricts itself to the ever-recurring cycles in the manner of ancient astrology.

We have seen that Tarnas adopts the general three-phase structure of the Hegelian dialectic for his big picture of the history of the Western mind and for the evolution of human consciousness. Along with his affirmation of the actualization of the principle of freedom, however, he also follows Jung in characterizing the goal, or telos, in more symbolic or archetypal terms as a *coniunctio oppositorum,* or sacred marriage (*hieros gamos*). This characterization is, of course, fully consistent with the third moment of Hegel's dialectic, which can be summed up with the notion of the *identity of identity and difference* (and it is in fact by the standard of such a complex identity that one can evaluate the concreteness or actuality of the principle of freedom). Though including such concepts as identity and difference, or universality and particularity, Tarnas's preferred terms for the *coniunctio,* or marriage, are *psyche* and *cosmos* and, as we have seen, at a more archetypal level, the *masculine* and *feminine* and the solar and lunar. In keeping with the Jungian view of development, the (re)union of these terms represents the overcoming of their alienation, especially during the modern period following the Copernican Revolution and the rise of the Cartesian–Newtonian paradigm. What I am suggesting here is that the sacred marriage, as envisioned by Tarnas, calls for a more explicit celebration of a second *coniunctio,* or *complexio oppositorum,* as Jung also defined the Self (and which he conceived as both the guiding spirit and goal of the individuation process, or the evolution of consciousness). In this case, the joining, or weaving together (*complexere*), of the more symbolical/archetypal and polycentric perspective of the ancient astrological perspective with the more conceptual/dialectical and monocentric perspective of the evolution of consciousness (as exemplified by Hegel, Teilhard, Steiner, and Aurobindo).

Though the full import of the role of astrology would have to await the publication of *Cosmos and Psyche,* the union (or at least the passionate engagement) of these two perspectives is already in evidence in *The Passion of the Western Mind,* in which both Hegel and Jung play central hermeneutical roles. At about the same time as the appearance of *Passion,* though neither of us knew of the other's work, I argued a more restricted and technical case for such a union in my *Individuation and the Absolute,* where I demonstrate that Jung's concept of the Self is both structurally and functionally equivalent to Hegel's concept of Spirit (*Geist*), or the Absolute. Because the union is complex, however, I concluded that the terms in relation will maintain a certain creative tension—it is a matter, after all, of the identity of identity *and* difference.

6. *Conclusion: The Spiral of Evolution*

It is a truism among those given to the spirit of mediation that the seeming contradiction between the arrow of time and the ever-recurring cycle is resolved in the image of the spiral. The challenge has been to describe exactly what the spiral of history or evolution would look like in the details. In the body of this text, I have outlined an approach to the last two millennia, at least, where such a spiral appears to be in evidence. To summarize: what I found was that the three-phase structure of the overall trajectory seems to be repeated fractally, at least once between the first and second phases of the larger arc, and several times between the second and third. The three main phases are 1) identity, the biblical worldview as Great Code; 2) difference, the Modern worldview (as secularization of the Great Code); and 3) an accelerated movement toward the identity of identity and difference, the Planetary worldview (a new identity in the making). The first turn of the spiral is traced by the movement from the life of the early, mythically embedded, Christian community, through the differentiation of consciousness effected by medieval scholasticism and the establishment of the worldly power of Christendom, to the birth of modernity with the Renaissance, the Copernican Revolution, and the Protestant Reformation, which together, despite their negation of the preceding medieval worldview, represent the forging of a new, more complex identity. The countercultural elements of this new identity are themselves negated with the Enlightenment and the emergence of the fully secularized (and materialistically oriented) Cartesian–Newtonian paradigm, which in turn generates a second, fractal third moment in the countercultural response of Romanticism and speculative Idealism (culminating in Hegel).

As we have seen, there have been three more ever-tightening turns of the spiral from then to the present. Despite the revolutionary developments in physics, psychology, spirituality, and the arts surrounding the threshold of the twentieth century, the previous two centuries have been dominated by what Van Baumer calls the "New Enlightenment" (which includes such movements as positivism, Marxism, Freudianism, and a general faith in the power of economic and technological progress). Without actually overturning the dominant trend, the sixties counterculture succeeded in constituting a distinct new identity while taking up, however unwittingly, many of the themes and much of the spirit of the earlier countercultural surge of the Romantic–Idealist era (especially with its organicist, participatory, and enchanted view of the cosmos). As we know, the counterculture was succeeded by a conservative entrenchment in politics and in culture generally. At least in the United States, one can discern a third turn of the spiral with the movement through the more optimistic and spiritually open years of the Clinton administration to the recent phase dominated by the neoconservative agenda. Encouraged by Tarnas's reading of approaching world transits, I have risked the prediction that we are on the threshold of another countercultural upsurge—though, of course, with mounting tensions and deepening complexity, everything is increasingly uncertain.

The more compelling indicators, which confirm the rate at which the spiral is tightening, are the currently unfolding world transits involving Pluto, Uranus, and Saturn, the configuration of which shows a clear resonance with the sixties counterculture (and earlier with the time of the French revolutionary upheavals immediately preceding the birth of the Romantic–Idealist era). Despite these illuminating correspondences, however, I am left with the following observations with respect to the union of the astrological with the Hegelian and Jungian perspectives: even in spiral form, the overall trajectory points to a kind of Omega point, or Singularity. What are the implications for our collective human relationship to the planetary archetypes as we draw ever nearer to this Singularity, or perhaps even pass through it to whatever lies on the other side? Early Christians, both Gnostic and orthodox, imagined that the singular salvific Event by which their time—and all time—was defined had the potential of freeing them from the fatalism of the stars. Two millennia later, we have the potential, at least, to feel more at home in a larger and more complex unfolding cosmos. Still, at the time of writing, a sixty-five-million-year geological era (the Cenozoic) is coming quickly to an end through the human-created Sixth Mass Extinction. The very biosphere as we have always known it is threatened with collapse, the resource base of the modern era is being rapidly depleted, and there are billions of human lives trapped in desperate suffering. The goal of fully actualized freedom on a planetary scale seems impossibly distant. While the archetypal–astrological perspective and that of the spiral of evolution I have traced in the preceding pages may illuminate the path immediately ahead, the Earth itself has become a "wanderer" (*planētēs*) in a sense scarcely imaginable by the ancients. Whether here on Earth or directed to the heavens, we see only as in a glass, darkly.

NOTES

Introduction

1. See Morin (1999) 5ff.
2. For links to the most comprehensive database on the Sixth Mass Extinction, see www.speciesalliance.org.
3. Berry (1999) 4ff.
4. In this connection, see the *Global Europe Anticipation Bulletin* (http://www.leap2020.eu/).
5. Tarnas (1993) 492.
6. In Abrams (1973) 251; my translation.

Chapter 1: The Monomyth

7. Campbell (2003) 28; italics in original. For a good cross-cultural collection of myths illustrating the universality of the *monomyth,* see David Adams Leeming's *Mythology: The Voyage of the Hero* (Oxford: Oxford University Press, 1998).
8. See Baring and Cashford (1993), *The Myth of the Goddess,* pp. 273ff.
9. See Jaynes (2000), *The Origins of Consciousness in the Breakdown of the Bi-Cameral Mind.*
10. Campbell (1977) 163.
11. Ibid.
12. Odyssey, 333.
13. Tolkien (1965) 385.

Chapter 2: A More Fundamental Pattern

14. See Kelly (1988).
15. Kelly (1993a).
16. Barfield (1988 [1965]) 182–183.
17. Wilber (1995).
18. Grof (1985 and 1988).

19. Washburn (1995).
20. Gebser (1986).
21. Tolkien (2000) no. 211, p. 283.
22. On the Christian and specifically Catholic dimensions of Tolkien's work, see especially Joseph Pearce's *Literary Giants, Literary Catholics* (Pearce, 2005) and Ralph Wood's *The Gospel According to Tolkien* (Wood, 2003).

Chapter 3: From Myth to History

23. See Wilber (1999c) 564.
24. Ibid., 686.
25. See Eliade (1987) 52, 62, 64, and 67ff.
26. Campbell (1988) 111.
27. Cobb (2002).
28. Morin (1999) 1.
29. Jaspers (1968) 24.
30. Ibid., 1.
31. McNeill and McNeill (2006) 322.
32. Ibid., 155.
33. Korotayev (2004) 86.
34. Ibid., 88.
35. Thompson (2001).
36. Kelly (1993a).
37. Quoted in ibid., 165.

Chapter 4: The Great Code

38. McNeill (1965); see also Morin (1999) 1–26.
39. Berger (1990).
40. Grosso (1995).
41. Thompson (1991) 255.
42. See Kelly (1993a).
43. Angus (1929) 94–95.
44. Ibid., 101.
45. Jung, C.W 9 ii: 78; see also Kelly (1993a) 168f.
46. Jaspers, 23.

179

47. See Grant (1996).
48. See Bala (2006).
49. Huff (2003).
50. Harrison (2001).
51. Kojève (1964) 301; my translation.
52. Ibid., 303.
53. Ibid., 304–305.

Chapter 5: Birthing the Modern

54. Löwith, 146.
55. See Grosso, 45.
56. Löwith, 158–159.
57. See Tarnas (1993) especially 200ff.
58. Merchant (1989) 103.
59. Cited in Merchant, 104.
60. Ibid, 115.
61. Cited in Torrance (1999) 726.
62. Yates (1991) 433.
63. White (1997) 131–132.
64. Hill (1991) 291.
65. See Hill (1991) 289.
66. Quoted in Hill (1991) 142.
67. Luther (1520) in Ross and McLaughlin, eds (1981), 722.
68. Barzun (2000) 270.
69. In Hill (1991) 107.
70. Ibid., 148.
71. Ibid., 206.
72. Ibid., 199.
73. Ibid., 296.
74. Ibid., 297.

Chapter 6: Triumph, Revolution, and Protest

75. Yates (1991) 452.
76. Ibid., 436.
77. Ibid. (1991) 453.
78. Ibid., 448.
79. Lessing (1778).
80. Quoted in Löwith, 93.
81. Abrams, 334
82. Ibid.
83. Gusdorf (1993) 317–318, my translation.
84. Ibid., 436–437, my translation.
85. Richards (2004) 11.

86. Schelling, quoted in Baumer (1978) 486.
87. Quoted in Gusdorf, 390, my translation.
88. Abrams, 190–191.
89. Hodgson in Hegel (1988) 45.

Chapter 7: From the New Enlightenment to the Counterculture

90. Baumer, 468.
91. From Marx's 1844 open letter to Arnold Ruge (Marx, 1844).
92. Bernard, in Baumer, 534.
93. Comte, in ibid., 525.
94. Ibid., 527.
95. Ibid., 525.
96. See Baumer, 463.
97. Bohm (2002).
98. Thompson (2004) 45–47.
99. Cited in Hanegraaff (1998) 448.
100. Ibid., 449.
101. Baumer (1978) 647.
102. Ibid., 651.
103. See ibid., 655.
104. Roszak (1969) 13.
105. Ibid., 8.
106. In Roszak (1969) 22.
107. Roszak, 66.
108. Morin (2008) 12.
109. See Richards, 17ff.
110. See Abrams; and Gusdorf.
111. See Godwin (1994).
112. Spretnak (1997) 176.

Chapter 8: A Tightening Spiral—A Widening Gyre

113. Jenkins (2006).
114. Thompson (1986) 181.
115. Thompson (1991) 256.
116. Tarnas (2005) 419.
117. Tarnas, ibid., 427.

Chapter 9: Lengthening Shadows

118. Spretnak (1997) 217.
119. Ibid., 41.

Chapter 10: Global Solidarity

120. See Monbiot (2004).
121. Hawken (2007) 24.
122. From the "Who We Are" page of the Bioneers website: www.bioneers.org.
123. Ibid.
124. Ibid.
125. Morin (1999) 141.
126. Ibid.
127. Ibid., 142.
128. Morin in Kelly (1999) 236.
129. Ibid.

Chapter 11: From the Soul of the World

130. For a discussion of the differences and continuity between "apocalyptic" and "millenarian" in the context of the New Age movement, see Hanegraaff, 98ff. In this context, I use the two terms more or less interchangeably.
131. Hanegraaf (1998) 12.
132. Ibid., 10.
133. Ibid., 67.
134. On the links between the mechanistic paradigm, environmental degradation, and the oppression of women, see Merchant (1983).
135. Bohm (2002) 14ff.
136. For an account of quantum nonlocality or entanglement, from the Einstein–Podolsky–Rosen thought experiment, though Bell's Theorem, to its experimental verification, see Herbert, 199f.
137. See Sheldrake (1989).
138. See Morin (1992).
139. See Wilber (1995) 414ff. and 651ff.
140. See Lovelock (1965, 1972 and 2007); and Harding (2006).
141. Quoted in Hanegraaff, 65.
142. On the history of transpersonal psychology and its precursors, see Rothberg and Kelly, (1998) 1ff.; and Kelly (2002).
143. Grof (1985).
144. See Grof (1988).
145. See Stevens (1998).
146. See Kelly, forthcoming.
147. See Wilber (1999b) 264ff.
148. Ibid., 153.
149. Wilber (1983) 314.
150. Abrams, 185.
151. For comprehensive treatments of Romanticism, see not only Abrams, but Richards; and Gusdorf.
152. Tarnas (2006) 456.
153. Ibid., 479.
154. Ibid., 468.
155. Ibid.

Chapter 12: First Light: Toward a Planetary Wisdom Culture

156. See Kelly (1993a) 175ff.
157. In Jung (1982) 280.
158. Jung, C.W. 11: 7.
159. Macy (1998) 25ff.
160. See Wilber (1999d) 122 and 132.
161. Quoted in Abrams, 215 and 267.
162. Kelly (1993a)
163. See Morin (1992).
164. Morin (1977) 80.
165. Ibid., 86.
166. Morin (1986) 419.
167. Swimme (1996) 85, 111, 112.
168. For a discussion of these two senses of participation in relation to Morin's paradigm of complexity, see my "Participation, Complexity, and the Study of Religion," in Ferrer and Sherman (2008 a).
169. Gardner, 25.
170. American Anthropological Association Statement on "Race"—(May 17, 1998). http://www.aaanet.org/stmts/racepp.htm [8/5/05].
171. Morin (1999) 40–41.
172. Quoted in Kelly (1993a) 184.

173. See www.guardian.
co.uk/climatechange/
story/0,12374,1517831,00.html
(August 8, 2005).

174. See www.panda.org/about_our_
earth/all_publications/living_
planet_report/ [July 5, 2009).

175. See www.massextinction.
net [7/5/09]. See also www.
speciesalliance.org.

176. Earth Charter, see www.earthchar-
ter.org .

177. Ferrer (2002) 146–7.

178. For an extended treatment of the
complex character of these and
other oppositions in the context of
religious studies, see Kelly (2008).

179. Cousins (1992) 10.

180. Thompson (2004) 76.

181. From Morin's response in Kelly,
Bohm, and Morin (1997) 235.

182. Hegel (2008 [1820]) 21.

Epilogue

183. Hawken, 5.

184. Korten (2006) 20.

185. Ibid.

186. See Martin (2006).

187. Morin (1999) 100.

188. Ibid., 149.

Appendix II

189. See Tarnas (2006) 489 and 492.

190. For Tarnas's position on this matter, see Tarnas (2006) 137.

References

Abrams, M. H. (1973). *Natural Supernaturalism: Tradition and Revolution in Romantic Literature*. New York: Norton.

Angus, S. (1929). *The Religious Quests of the Graeco-Roman World: A Study in the Historical Background of Early Christianity*. New York: Scribner's.

Armstrong, K. (2006). *The Great Transformation: The Beginning of Our Religious Traditions*. New York: Knopf.

Aurobindo (1951). *The Life Divine*. New York: Dutton.

Bala, A. (2006). *The Dialogue of Civilizations in the Birth of Modern Science*. Hampshire, UK: Palgrave Macmillan.

Barfield, O. (1988[1965]). *Saving the Appearances: A Study in Idolatry*. Lebanon, NH: Wesleyan University Press.

Baring, A, and J. Cashford (1993). *The Myth of the Goddess*. London: Arkana.

Barnhart, B. (2007). *The Future of Wisdom: Toward a Rebirth of Sapiential Christianity*. New York: Continuum.

Barzun, J. (2000). *From Dawn to Decadence: 500 Years of Western Cultural Life, 1500 to the Present*. New York: HarperCollins.

Baumer, F. Le Van. (1978). *Main Currents of Western Thought: Readings in Western Europe Intellectual History from The Middle Ages to the Present*. New Haven, CT: Yale University Press.

Berger, P. (1990). *The Sacred Canopy: Elements of a Sociological Theory of Religion*. New York: Anchor.

Berry, T. (1999). *The Great Work: Our Way into the Future*. New York: Random House.

Bohm, D. (2002). *Wholeness and the Implicate Order*. New York: Routledge Classics.

Blavatsky, H. P. (1972). *The Key to Theosophy*. Wheaton, IL: Theosophical University Press.

Campbell, J. (1977). *The Masks of God: Occidental Mythology*. Harmondworth, UK: Penguin.

———(1988). *The Way of the Seeded Earth, Part 1: The Sacrifice (Historical Atlas of World Mythology)*. New York: Harper & Row.

———(2003). *The Hero with a Thousand Faces*. Princeton, NJ: Princeton University Press.

Charet, F. X. (1993). *Spiritualism and the Foundations of C. G. Jung's Psychology*. Albany, NY: SUNY Press.

Cobb, J. (2002). "Constructive Postmodernism." In Religion Online, www.religion-online.org/showarticle.asp?title=2220 (retrieved: 7/3/09).

Cousins, E. H. (1992). *Christ of the 21st Century*. Rockport, MA: Element.

Cusa, N. (1997). *Nicholas of Cusa: Selected Spiritual Writings* (H. Lawrence Bond, tr.), New York: Paulist.

Earth Charter: www.earthcharter.org/files/charter/charter.pdf (retrieved: 6/20/07).

Eliade, M. (1987). *The Sacred and The Profane: The Nature of Religion*. Orlando: Harcourt.

Ferrer, J. (2002). *Revisioning Transpersonal Theory: A Participatory View of Human Spirituality*. Albany, NY: SUNY Press.

———(2008). Ferrer, J., and J. Sherman. eds. *The Participatory Turn: Spirituality, Mysticism, Religious Studies*. Albany, NY: SUNY Press.

Gardner, J. (2003). *Biocosm: The New Scientific Theory of Evolution*. Makawao, Maui: Inner Ocean Books.

Gebser, J. (1986). *The Ever-present Origin*. Athens OH: Ohio University Press.

Godwin, J. (1994). *The Theosophical Enlightenment*. Albany, NY: SUNY Press.

Grant, E. (1996). *The Foundations of Modern Science in the Middle Ages: Their Religious, Institutional, and Intellectual Contexts*. Cambridge, UK: Cambridge University Press.

Grof, S. (1985). *Beyond the Brain: Birth, Death, and Transcendence in Psychotherapy*. Albany, NY: SUNY Press.

———(1988). *The Adventure of Self-Discovery*. Albany, NY: SUNY Press.

Grosso, M. (1995). *The Millennium Myth: Love and Death at the End of Time*. Wheaton, IL.: Quest Books.

Gusdorf, G. (1993). *Le romantism.* (deux volumes: *Le savoir romantique* et *l'homme et la nature*). Paris: Editions Payot et Rivages.

Hanegraaff, W. (1998). *New Age Religion and Western Culture: Esotericism in the Mirror of Secular Thought*. Albany, NY: SUNY Press.

Harding, S. (2006). *Animate Earth: Science, Intuition, and Gaia*. White River Junction, VT: Chelsea Green.

Harrison, P. (2001). *The Bible, Protestantism, and the Rise of Natural Science*. Cambridge, UK: Cambridge University Press.

Hawken, P. (2007). *Blessed Unrest: How the Largest Movement in the World Came into Being and Why No One Saw It Coming*. New York: Viking.

Hegel, G. W. F. (1956). *Philosophy of History* (tr. J. Sibree). New York: Dover.

———(1975). *Hegel's Logic* (Part I of the *Encyclopaedia of the Philosophical Sciences* (tr. W. Wallace). New York: Oxford University Press.

———(1978). *The Difference between the Fichtean and Schellingian Systems of Philosophy* (tr. J. P. Surber). Atascadero, CA: Ridgeview Publishing.

———(1981). *The Phenomenology of Spirit* (tr. A. V. Miller). Oxford, UK: Oxford University Press.

———(1988). *Lectures on the Philosophy of Religion. One volume edition* (ed. P. C. Hodgson). Berkeley: University of California Press.

———(2008[1820]). *Philosophy of Right*. New York: Cosimo Classics.

Herbert, N. (1985). *Quantum Reality: Beyond the New Physics. An Excursion into Metaphysics*. New York: Anchor.

Hill, C. (1991). *The World Turned Upside Down: Radical Ideas during the English Revolution*. London: Penguin.

Huff, T. E. (2003). *The Rise of Early Modern Science: Islam, China and the West*. Cambridge, UK: Cambridge University Press.

Jaspers, K. (1968). *The Origin and Goal of History*. New Haven: Yale University Press.

Jaynes, J. (2000). *The Origins of Consciousness in the Breakdown of the Bi-cameral Mind*. New York: Houghton Mifflin.

Jenkins, P. (2006). *Decade of Nightmares: The End of the Sixties and the Making of Eighties America*. Oxford, UK: Oxford University Press.

Joyce, J. *Finnegans Wake*. (1999) New York: Penguin.

Jung. C. G. (1953–1979). *The Collected Works of C. G. Jung*. (tr. R. F. C. Hull). Boston: Princeton University Press.

———Volume 8: *The Structure and Dynamics of the Psyche*.

———Volume 11: *Psychology Religion: West and East*.

———Volume 9 II. *Aion: Researches into the Phenomenology of the Self*.

———(1982). *Memories, Dreams, Reflections*. Glasgow: Collins Fount.

Kaplan, R. (2001). *The Coming Anarchy*. New York: Vintage Books.

Kelly, S. (1988). "Hegel and Morin: the Science of Wisdom and the Wisdom of the New Science," in *The Owl of Minerva: The Biannual Journal of the Hegel Society of America*. 20, 1. pp. 51–67.

———(1990 a). (with David Bohm) "Dialogue on Science, Society, and the Generative Order," in *Zygon: Journal of Religion and Science*. 25,4. pp. 449–467.

———(1991). "The Prodigal Soul: Religious Studies and the Advent of Transpersonal Psychology," in *Religious Studies: Issues, Prospects, and Proposals*. Atlanta: Scholars Press. pp. 429–441.

———(1992). "Beyond Materialism and Idealism," in *Idealistic Studies*. xxii, 1. pp. 28–38.

———(1993 a) *Individuation and the Absolute: Hegel, Jung, and the Path toward Wholeness*. Mahwah, NJ: Paulist.

———1993 b). "The Rebirth of Wisdom: Reflections on Richard Tarnas's *The Passion of the Western Mind*," in *San Francisco Jung Institute Library Journal*. 11, 4. pp. 33–44.

———(1993 c). "The Great Mother/Goddess and the Psychogenesis of Patriarchal Consciousness," in *Journal of Dharma*. xviii, no. 2. pp. 114–124.

———(1997) (with David Bohm and Edgar Morin). "Order, Disorder, and the Absolute: An Experiment in Dialogue," in *World Futures*, Vol. 46. pp. 223–237.

———(1998). (coeditor, with Donald Rothberg): *Ken Wilber in Dialogue: Conversations with Leading Transpersonal Thinkers*. Wheaton, IL: Quest Books.

———(1999) "From the Complexity of Consciousness to the Consciousness of Complexity," in *ISSS 1999: Proceedings of the International Society of the Systems Sciences*. Pine Grove, CA: Asilomar.

———(2000). "Transpersonal Psychology and the Paradigm of Complexity," in *Integralis: Journal of Integral Consciousness, Culture and Science*. vol. 1., no. 0 (premier online issue at www.integralage.org).

———(2002). "Space, Time, and Spirit: The Analogical Imagination and the Evolution of Transpersonal Theory," in *Journal of Transpersonal Psychology*. vol. 34, no. 2, 2002, pp. 73–100.

———(2008 a). "Participation, Complexity, and the Study of Religion," in Ferrer, J., and J. Sherman, eds. *The Participatory Turn: Spirituality, Mysticism, Religious Studies*. Albany, NY: SUNY Press.

———(2008 b). "Integral Time and the Varieties of Post-Mortem Survival," in *Integral Review*. vol. 4, no. 1, pp. 5–30. (integral-review.org/index.asp)

———(forthcoming). "James, Grof, and the Varieties of Perinatal Experience," in a *Feschrift for Stanislav Grof* (ed. Richard Tarnas).

Kojève, A. (1964). "L'origine chrétienne de la science moderne," in *Mélanges Alexandre Koyré: L'aventure de l'esprit*. Paris: Hermann. pp. 295–306.

Korotayev, A. (2004). *World Religions and Social Evolution of the Old World Oikumene Civilizations: A Cross-cultural Perspective.* Lampeter, Ceredigion, UK: Edwin Mellen.

Korten, D. (2006). *The Great Turning: From Empire to Earth Community.* Bloomfield, CT: Kumarian Press.

Koyré. A. (1957). *From the Closed World to the Infinite Universe.* Baltimore: Johns Hopkins University Press.

Leeming, D. A. (1998). *Mythology: The Voyage of the Hero.* Oxford, UK: Oxford University Press.

Lessing, G. E. (1778) "The Education of the Human Race," in *Literary and Philosophical Essays: French, German, and Italian* (with introductions and notes). New York: Collier [c.1910] Harvard Classics, 32.

Lovelock, L. (1965). "A Physical Basis for Life Detection Experiments," in *Nature* 207 (7): 568–570.

———(1972). "Gaia as Seen through the Atmosphere," in *Atmospheric Environment* 6 (8): 579–580.

———(2007). *The Revenge of Gaia: Earth's Climate Crisis and the Fate of Humanity.* New York: Basic Books.

Löwith, K. (1949). *Meaning in History.* Chicago: University of Chicago Press.

McNeill, J. R., and W. McNeill. (2006). *The Human Web: A Bird's-eye View of World History.* New York: Norton.

McNeill, W. H. (1965). *The Rise of the West: A History of the Human Community.* New York: New American Library.

Macy, J. (1998). *Coming Back to Life: Practices to Reconnect our Lives, our World.* Gabriola Island, BC: New Society Publishers.

Martin, J. (2006). *The Meaning of the 21st Century: A Vital Blueprint for Ensuring Our Future.* New York: Riverhead Books.

Marx, K. (1844). "For a Ruthless Criticism of Everything Existing," in *The Marx–Engels Reader.* Ed. R. Tucker. New York: Norton, 1972. pp. 7-10.

Merchant, C. (1989). *The Death of Nature: Women, Ecology, and the Scientific Revolution.* San Francisco: HarperSanfrancisco.

Monbiot, G. (2004). *Manifesto for a New World Order.* New York: New Press.

Morin, E. (1977). *La méthode 1: La nature de la nature.* Paris: Seuil.

———. (1986). *La méthode 3: La connaissance de la connaissance.* Paris: Seuil

———(1992). "From the Concept of System to the Paradigm of Complexity," in the *Journal of Social and Evolutionary Systems.* 15, 4. , pp. 371-385 (tr. S. Kelly).

———(1999) (with A. B. Kern) *Homeland Earth: A Manifesto for the New Millennium* (tr. S. Kelly and R. Lapointe). Kreskill, NJ: Hampton Press.

———(2004). *La Méthode 6: Éthique.* Paris: Seuil.

———(2008). *California Journal.* Brighton, UK: Sussex Academic Press.

Odyssey of Homer (1991) (tr. Richard Lattimore). New York: Harper Perennial.

Pearce, J. (2005). *Literary Giants, Literary Catholics.* Fort Collins, CO: Ignatius Press.

Ray, P. H., and S. R. Anderson. (2001). *The Cultural Creatives: How 50 Million People Are Changing the World.* New York: Three Rivers.

Richards, R. J. (2004). *The Romantic Conception of Life: Science and Philosophy in the Age of Goethe.* Chicago: University of Chicago Press.

Reeves, M., and Gould, W. (1987). *Joachim of Fiore and the Myth of the Eternal Evangel in the Ninetheenth Century*. Oxford, UK: Clarendon Books.

Ross, J. B., and McLaughlin, M. M., eds. (1981). *The Portable Renaissance Reader*. New York: Penguin.

Rothberg, D., and Kelly, S., eds. (1998). *Ken Wilber in Dialogue: Conversations with Leading Transpersonal Thinkers*. Wheaton, IL: Quest Books.

Schelling, F. W. J. (1984). *Bruno, or, On the Natural and the Divine Principle of Things*. Albany, NY: SUNY Press.

Sheldrake, R. (1989). *The Presence of the Past: Morphic Resonance and the Habits of Nature*. New York: Vintage Books.

Spretnak, C. (1997). *The Resurgence of the Real: Body, Nature, and Place in a Hypermodern World*. Reading, MA: Addison-Wesley.

Stevens, J. (1998). *Storming Heaven: LSD and the American Dream*. Grove.

Swimme, B. (1996). *The Hidden Heart of the Cosmos: Humanity and the New Story*. Maryknoll, NY: Orbis.

Tarnas, R. (1991). *The Passion of the Western Mind: Understanding the Ideas that Have Shaped Our Worldview*. New York: Ballantime.

————(2006). *Cosmos and Psyche: Intimations of a New World View*. New York: Viking Penguin.

Thompson, W. I. (1974). *Passages about Earth: An Exploration of the New Planetary Culture*. New York: Harper & Row.

————(1986). *Pacific Shift*. San Francisco: Sierra Club.

————(1985/1986). "It's Already Begun: The Planetary Age Is an Unacknowledged Daily Reality." *In Context*, no. 12, winter 1985/86, p. 26 (retrieved, 6/24/09, from www.context.org/ICLIB/IC12/Thompson.htm).

————(1991). *Gaia 2: Emergence: The New Science of Becoming*. Hudson, NY: Lindisfarne.

————(2001). *Transforming History: A Curriculum for Cultural Evolution*. Great Barrington, MA: Lindisfarne. (NOTE: the 2001 edition was withdrawn; a thoroughly revised edition is *Transforming History: A New Curriculum for a Planetary Culture*. Lindisfarne, 2009).

————(2004). *Self and Society: Studies in the Evolution of Culture*. Exeter, UK: Imprint Academic.

Tolkien, J. R. R. (1965). *The Lord of the Rings: The Fellowship of the Ring; The Two Towers; The Return of the King*. New York: Ballantine Books.

————(2000). *The Letters of J. R. R. Tolkien* (ed. Christopher Tolkien). New York: Houghton Mifflin.

Torrance, R. M. ed. (1999) *Encompassing Nature: A Source Book—Nature and Culture from Ancient Times to the Modern World*. Washington, DC: Counterpoint.

Washburn, M. (1995). *The Ego and the Dynamic Ground: A Transpersonal Theory of Human Development*. Albany, NY: SUNY Press.

Weber, E. (1999). *Apocalypses: Prophecies, Cults, and Millennial Beliefs through the Ages*. Cambridge, MA: Harvard University Press.

White, M. (1997). *Isaac Newton: The Last Sorcerer*. New York: Helix Press.

Whitehead, A. N. (1978). *Process and Reality. Corrected Edition*. New York: Macmillan.

Wilber, K. (1983). *Up from Eden: A Transpersonal View of Human Evolution.* Boston: Shambhala.

———(1995). *Sex, Ecology, Spirituality: The Spirit of Evolution.* Boston: Shambhala.

———(1999 a). *Up from Eden,* in *The Collected Works of Ken Wilber,* vol. 2. Boston: Shambhala.

———(1999 b). *Eye to Eye,* in *The Collected Works of Ken Wilber,* vol. 3. Boston: Shambhala.

———(1999 c). *Integral Psychology,* in *The Collected Works of Ken Wilbe r,*vol. 4. Boston: Shambhala.

———(1999 d). *The Marriage of Sense and Soul,* in *The Collected Works of Ken Wilber,* vol. 8. Boston: Shambhala.

———(2006). *Integral Spirituality: A Startling New Role for Religion in the Modern and Postmodern World.* Boston: Shambhala.

Wood, R. (2003). *The Gospel According to Tolkien: Visions of the Kingdom in Middle-earth.* Louisville, KY: Westminster John Knox.

Worster, D. (1994). *Nature's Economy: A History of Ecological Ideas.* New York: Cambridge University Press.

Yates, F. (1991). *Giordano Bruno and the Hermetic Tradition.* Chicago: University of Chicago Press.

INDEX

ACKNOWLEDGMENTS

I WISH to express my gratitude to my beloved friends and colleagues for their inspiration, encouragement, and helpful responses to earlier drafts of this project: especially to Jenny Connolly, Robert McDermott, and Richard Tarnas, with whom I have been in ongoing dialogue and who gave very helpful feedback on my ideas as they evolved. Richard in particular gave the penultimate draft a very close reading, resulting in many and significant improvements. Gratitude also to Edgar Morin and Joanna Macy, two wise elders and friends, whose insights and whose lives are beacons to me in my aspiration to wisdom; to my loving siblings, Ken, Jo Ann, and Kristine; and finally to Cynthia Morrow, Brian Swimme, David Ulansey, Charlene Spretnak, Jorge Ferrer, Michael Murphy, Frank Poletti, Sam Mickey, Christof Papajewski, Ève Brière; and to my students in the Philosophy, Cosmology, and Consciousness program at the California Institute of Integral Studies where I teach (and continue to learn)—all of whom have helped me in very particular ways.

www.ingramcontent.com/pod-product-compliance
Lightning Source LLC
Chambersburg PA
CBHW022127080426
42734CB00006B/258